Introduction to Intellectual Property Rights

The Author

Phundan Singh [b. 1946] hails from a reputed agricultural family of Western Uttar Pradesh [Village-Sakauti, Tehsil-Mawana, District-Meerut]. He graduated in Agriculture from Agra University, Agra and obtained his Master and Doctorate degrees from Kanpur University, Kanpur. He has throughout a brilliant academic record. His area of specialization is Plant Breeding and Genetics.

Dr. P. Singh has total experience of 44 years in the field of Genetics and Plant Breeding. He has served the Central Institute for Cotton Research, Nagpur for about 32 years in various capacities such as Scientist, Senior Scientist, Principal Scientist, Head of Division and Director. He has 300 publications to his credit. He has participated and presented 80 research papers in various National and International Seminars, Symposia and Conferences. He has authored 52 books on Genetics and Plant Breeding. He has also authored 18 technical bulletins on various aspects of cotton and contributed 14 chapters in various books.

Dr P. Singh is a member of several Scientific Societies, referee of different journals, Expert member of Selection Committees and M.Sc. and Ph.D. Examiner in different Agricultural Universities. The Institute was recognized as the referral Laboratory for detection of *Bt.* gene and received the best Annual Report Award when he was the Director. He has visited several countries such as Russia, Belarus, Ukraine, Crimea, Uzbekistan, Canada, USA, England, Austria and Germany.

Introduction to Intellectual Property Rights

Phundan Singh

Former Director
Central Institute for Cotton Research,
Nagpur – 440 010
Maharashtra, India

2016

Daya Publishing House®

A Division of

Astral International Pvt. Ltd.

New Delhi – 110 002

Publisher's Note:

Every possible effort has been made to ensure that the information contained in this book is accurate at the time of going to press, and the publisher and author cannot accept responsibility for any errors or omissions, however caused. No responsibility for loss or damage occasioned to any person acting, or refraining from action, as a result of the material in this publication can be accepted by the editor, the publisher or the author. The Publisher is not associated with any product or vendor mentioned in the book. The contents of this work are intended to further general scientific research, understanding and discussion only. Readers should consult with a specialist where appropriate.

Every effort has been made to trace the owners of copyright material used in this book, if any. The author and the publisher will be grateful for any omission brought to their notice for acknowledgement in the future editions of the book.

Cataloging in Publication Data--DK
Courtesy: D.K. Agencies (P) Ltd. <docinfo@dkagencies.com>

Singh, Phundan, 1946- author.
Introduction to intellectual property rights / Phundan Singh.
pages cm
Includes bibliographical references.

ISBN 978-93-5130-969-7 (International Edition)

1. Intellectual property. 2. Intellectual property--India. I. Title.

K1401.S56 2016 DDC 346.048 23

Published by : **Daya Publishing House®**
A Division of
Astral International Pvt. Ltd.
– ISO 9001:2008 Certified Company –
4760-61/23, Ansari Road, Darya Ganj
New Delhi-110 002
Ph. 011-43549197, 23278134
E-mail: info@astralint.com
Website: www.astralint.com

Laser Typesetting : **Classic Computer Services,** Delhi - 110 035

— Dedicated to —

My beloved Late Parents:
Mrs. Asharphi Singh and
Shri Mangat Singh

Preface

The subject of Intellectual Property Rights is gaining increasing importance these days. There are several books on this subject both of foreign and Indian origin. However, the complete information is not available from any single source and one has to go through several voluminous books to get the complete information. The present book has been written to fill up this gap.

This book provides comprehensive information on the subject in one compact volume comprising 20 chapters. The book has been divided into two sections. Section 1 deals with Primary Rights consisting of 12 chapters [Chapter 1 to Chapter 12]. Section 2 deals with Sui-generic Rights comprising 8 Chapters [Chapter 13 to Chapter 20]. The information contained in this book has been gathered from various published sources and Internet websites. Attempts have been made to provide latest information even then some valuable information might have been missed.

The Glossary of important terms used in relation to Intellectual Property Rights and Farmers Rights has been provided at the end for ready reference. Important References have been listed at the end which would be useful in getting further details.

I could not spare adequate time for my wife, grand children and other family members during writing of this book for which I feel guilty. The cooperation extended by my wife Mrs. Jaswanti Singh during preparation of this book is praise worthy.

Hope this Volume would be useful to the students, researchers, Plant Breeders, Design developers, artists, musicians, private seed companies, etc. Constructive suggestions of readers are welcome for further improvement of this book.

Phundan Singh

Contents

Section II: Sui-Generis IP Rights

Section I

Primary IP Rights

Chapter 1

Outlines of IPR

Introduction

The property refers to wealth or valuable things earned by a person. It is estimated in terms of land, house, garden, industries, animals, gold, silver, diamond, money etc. accumulated by a person. Each person wants to earn property for four main purposes as given below:

(i) To get a better or higher status in the society, because of a person is measured in terms of property accumulated by him or her. On the basis of property, the person is called as rich, middle class or poor.

(ii) To lead a better quality of life in the society.

(iii) To get a name and fame in the society. In the society, rich people get more respect than others.

(iv) For better financial security of life for himself and for his children.

Types of Property

The various valuable belongings of a person are broadly classified into two groups, *viz.* (i) immovable property, and (ii) movable property. These are defined below:

Immovable Property

It refers to fixed types of properties which can not move from one place to other. Such properties include land, buildings [house, flat, bungalow, farm house] and gardens. Such properties are protected by laws of land in each country.

Movable Property

It refers to the property which can be easily shifted from one place to other. Such property includes animals, farm machines, furniture, fixtures, gold, silver, diamond, and money. Such properties are also protected by public laws.

Intellectual Property

There is third type of property what we call intellectual property. The product, process/idea which is outcome of the brain of a person and can be used on commercial scale for benefit of human kind is called intellectual property. In other words, Intellectual property refers to creations of the mind: inventions, literary and artistic works, and symbols, names, images, and designs used on commercial scale. The main features of intellectual property are given below:

1. It is measured in terms of new ideas, processes, products, inventions and innovations developed by a person.
2. It requires lot of intellectual inputs in terms of thinking, planning and fine tuning of new ideas/products/processes etc.
3. It requires considerable amount of funds and other resources to develop new products/processes.
4. The main problem with intellectual property is that it can be copied, reproduced and used by others resulting in loss of inventor. Hence protection of intellectual property is essential so that the inventor can derive maximum benefits from his invention.

The rights to intangible property that is the product of the human intellect are referred to as intellectual property rights. Intellectual property may be protected by copyright, trademark or patent. The holder of intellectual property rights is usually the person or persons who developed the product or the organization that funded it.

Forms of Intellectual Property Rights

The Intellectual Property Rights are broadly divided into two groups, *viz.*, Primary Rights and Sui-generis Rights. These are briefly discussed below:

1. Primary Rights

Primary rights include,Copyrights, Patents, Trademarks, Trade Secret, Trade names, Domain names, Industrial Designs,Geographical Indications etc.

2. Sui-generis Rights

Sui-generis refers to things of their own kind or things with unique characteristics. Such rights include Database Rights, Mask work, Plant Breeders Rights, Traditional Knowledge, Moral Rights and Supplementary Protection Certificate.

Primary Rights

1. Copyrights

Copyrights refer to the legal right granted to an author, composer, playwright, publisher, or distributor to exclusive publication, production, sale, or distribution of a literary, musical, dramatic, or artistic work is called copyright. Main features of copyrights are briefly presented below:

Table 1.1: Forms of Intellectual Property Rights

Sl.No.	Primary Rights	Sl.No.	Sui-generis Rights
1.	Copyrights	1.	Database Rights
2.	Patents	2.	Mask work
3.	Trademarks	3.	Plant Breeders'Rights
4.	Industrial Design Rights	4.	Farmers'Rights
5.	Utility Models	5.	Moral Rights
6.	Geographical Indication	6.	Supplementary Protection Certificate
7.	Trade Secret	7.	Indigenous Intellectual Property
8.	Related Rights		
9.	Trade Names		
10.	Domain Names		

(i) It provides protection for a specific duration.

(ii) It is applicable in all countries.

(iii) The copyright holder has the rights to authorize others to use the protected work.

(iv) It is applicable to books, movies, music, paintings, photographs and software.

(v) The information cannot be reproduced as such without written permission. However, the information or idea can be used by any one.

2. Patent

Patent is an official document which grants sole rights to the inventor for manufacturing and marketing his product/process/invention to derive benefits. Main points are as follows:

(i) It is a. legal document issued by patent office to the inventor.

(ii) It grants exclusive rights to the inventor for the production, sale and profit from the invention of a product or process.

(iii) It is issued for a specific period of time.

(iv) Patents grant the right to prevent others from copying the invention.

(v) Many audio and video technologies are covered by patents.

3. Trademark

A trademark or trade mark is a distinctive sign of some kind which is used by an individual, business organization or other legal entity to uniquely identify the source of its products and/or services to consumers, and to distinguish its products or services from those of other entities. Main features of trademarks are briefly presented below:

(i) A trademark can be a word, name, symbol, device or mark which is use to identify and distinguish the goods or services of one company fro goods or services of another.

(ii) It is the name of a product made by a particular person or company.

(iii) The period of protection for a trademark varies, but can generally b renewed indefinitely.

4. Industrial Design

An industrial design is the ornamental or aesthetic aspect of an article produce by industry or handicraft. Main points about industrial designs are given below:

(i) Industrial designs include any original shape, picture, or some combinatio applied to a useful article of manufacture.

(ii) Registration is essential to get protection.

(iii) Registration can be done from 5 to 15 years.

(iv) The owner has the right to transfer right to any other person.

(v) Protection is valid only in the country where the design has been registered

5. Utility Models

A utility model is an exclusive right granted for an invention, which allows th right holder to prevent others from commercially using the protected invention without his authorization, for a limited period of time. The important points abou utility model are listed below:

(i) Utility models are also known as petty patents, petty inventions, petty innovations, utility innovations, minor patents and small patent.

(ii) The rights conferred by utility models are similar to those granted by patent laws but has a shorter term.

(iii) They are registered with National Patent Office.

(iv) Registration is required to get protection.

6. Geographical Indication

A geographical indication is a sign used on goods that have a specific geographical origin and often possess qualities or a reputation that are due to that place of origin. Main points related to geographical indications are as follows:

(i) Geographical Indication may be an agricultural, natural or manufactured goods or product.

(ii) About 170 geographical indications have been registered by Lisbon Agreement members up to 1997. Now this number would much more.

(iii) So far geographical indications have been registered in some advanced countries because they have developed the system of protecting GI.

(iv) In the past, wines,spirits, cheeses, tobacco, which account for 88 per cent, have been registered as GI. Wines and spirits account for almost 71 per cent of all registrations.

(v) In addition to wines and spirits, there is need for protection of agricultural products as GI. For this there is need to establish a multilateral system of identification and registration of GI.

7. Trade Secret

A trade secret is anything [a formula, process, method, mechanism, tool, pattern or device] which the disclosing party desires to keep secret. The main points about trade secret are briefly presented as follows:

(i) There is no specific period for trade secret. It may continue life long or for generations together.

(ii) There is no need of registration for trade secret.

(iii) It does not provide opportunity to others for improvement of innovation.

(iv) It is not applicable to books, equipments, plant varieties, designs which are openly used.

8. Related Rights

Related rights are rights which are similar to authors' right but which are not connected with the actual author of the work Both authors' rights and related rights are copyrights. Main points about related rights are given below:

1. Related rights are similar to copyrights.
2. Related rights come under primary property rights.
3. Related rights are also known as neighboring rights.
4. Related Rights are covered by Rome Convention, 1961.

9. Trade Names

A trade name is the name under which a product is commercially known. The commercial name by which a chemical is known is called trade name. One chemical may have a variety of trade names depending on the manufacturers or distributors involved Main points related to trade name are given below:

(i) It is also known as business name, trading name, assumed name, brand name and corporate name.

(ii) It may or may not be registered.

(iii) It distinguishes a particular business from others.

(iv) It can be exclusive or non-exclusive.

(v) It is generally used on letter heads and bank accounts.

(vi) It is the name given to a particular substance by each manufacturing company.

(vii) It is the name under which a company conducts its business.

It is s the business name of the person or organization making and/or selling a particular product. "General Motors, Inc" is the "trade name" of the company.

10. Domain Names

Domain refers to the unique name that identifies an Internet site. Domain name have two or more parts separated by dots. For example www.ask-edi.com or **www bcbsks.com** The term **domain name** has multiple related meanings:

(i) It identifies a web site on the internet (*e.g.* company.com); also referre to as a URL.[Uniform Resource Locator].

(ii) It allows reference to Internet sites without knowing the true numerica address.

(iii) It refers to the official name of a computer connected to the Internet. (*e.g* "www.microsoft.com").

(iv) It is the unique name of a computer on the Internet that distinguishes i from other systems around the world.

They are sometimes colloquially (and incorrectly) referred to by marketers a "web addresses".

Advantages and Disadvantages of IPR

Intellectual Property Rights refer to the legal rights provided to an invento to derive economic benefits from his invention/innovation. The main advantage of IPR are as follows:

(i) It promotes healthy competition for invention/innovation among the intellectuals.

(ii) It helps in improving the quality of the product.

(iii) It makes available new ideas/technologies to different countries.

(iv) It leads to faster development of industries/organizations engaged in research and development work.

There are some disadvantages of Intellectual Property Rights which are briefly presented below.

(i) The procedure of registration, particularly of patents, is very lengthy.

(ii) It involves lot of money transaction in registration, renewal and licensing.

(iii) It invites lots of court cases due to infringements.

(iv) It may lead to monopoly of right holders, etc.

Summary

The property refers to wealth or valuable things earned by a person. It is estimated in terms of land, house, garden, industries, animals, gold, silver, diamond, money etc. accumulated by a person. The various valuable belongings of a person are broadly classified into two groups, *viz.*, 1. immovable property, and 2. movable property.

Immovable Property refers to fixed types of properties which can not move from one place to other. Such properties include land, buildings [house, flat, bungalow, farm house] and gardens. Such properties are protected by laws of land in each country.

Movable Property refers to the property which can be easily shifted from one place to other. Such property includes animals, farm machines, furniture, fixtures, gold, silver, diamond, and money. Such properties are also protected by public laws.

There is third type of property what we call intellectual property. The product/process/idea which is outcome of the brain of a person and can be used on commercial scale for benefit of human kind is called intellectual property.

The Intellectual Property Rights are broadly divided into two groups, *viz.*, Primary Rights and Sui-generis Rights. Primary rights include,Copyrights, Patents, Trademarks, Industrial designs rights, utility models, geographical indication, trade secret, related rights, trade name and domain name.

A brief account of primary property rights has been presented. Sui-generic rights have been discussed in the chapter 13. Advantages and disadvantages of IPR have been discussed.

Glossary

Copyrights: The exclusive rights to reproduce, sell and distribute a work, prepare derivative works and display the work publicly.

Domain Name: The names and words that companies designate for their registered Internet Web site addresses,

Exclusive Rights: The legal privileges that only a copryight owner has with respect to their copyrighted work

Geographical Indication: A geographical indication is a sign used on goods that have a specific geographical origin and often possess qualities or a reputation that are due to that place of origin.

Immovable Property: Fixed types of properties which can not move from one place to other. Such properties include land, buildings [house, flat, bungalow, farm house] and gardens.

Industrial Design: Any original shape, picture, or some combination applied to a useful article of manufacture. In other words, it refers tothe design of the mass-produced products of our everyday environment, from sinks and furniture to computers.

Intellectual Property Rights: The legal rights provided to an inventor to derive economic benefits from his invention/innovation

Intellectual Property: The product/process/idea which is outcome of the brain of a person and can be used on commercial scale for benefit of human kind.

Movable Property: The property which can be easily shifted from one place to other. Such property includes animals, farm machines, furnitures, fixtures, gold, silver, diamond, and money.

Patent: A document granting an inventor sole rights to an invention. It is an officia document which grants sole rights to the inventor for manufacturing anc marketing his product/process/invention to derive benefits.

Property: The wealth or valuable things earned by a person

Trade Name: The name under which a company conducts business, or by whicl its business, product/service are identified. It may or may not be registerec as a trademark

Trade Secret: A formula or process or device used in business that is not publishec or divulged which gives an advantage over competitors

Trademark: A trademark can be a word, name, symbol, device or mark which is used to identify and distinguish the goods or services of one company from goods or services of another.

Questions

1. **Define property. Describe briefly different type of properties.**

2. **What do the following terms signify?**

   a. Immovable property
   b. Movable property
   c. Intellectual property
   d. Sui-generis rights

3. **Define the following terms:**

   a. Copyright
   b. Patent
   c. Trademarks
   d. Trade Secret
   e. Industrial design
   f. Geographical Indication

4. **Differentiate between the following:**

   a. Movable and immovable property
   b. Copyright and patent
   c. Trademark and trade Secret
   d. Trade Name and Domain Name

5. **What is Intellectual Property? List different forms of Intellectual Properties and describe any one in detail.**

6. **Define exclusive rights. Explain briefly various types of primary rights?**

7. **Describe briefly advantages and disadvantages of Intellectual Property Rights.**

8. **Describe briefly the merits and demerits of copyrights, patents, trademark and trade secret.**

9. **Write short notes on any two of the following:**

   (i) Utility Models
   (ii) Related Rights
   (iii) Trade Name
   (iv) Domain Name

0. **Give in brief comparison of copyrights and patents.**
1. **Describe briefly trademarks and trade secret. Give their merits and demerits.**
2. **Describe briefly utility models and related rights.**
3. **Describe briefly trade names and domain names.**
4. **Write short notes on the following:**
   - (i) Indigenous Knowledge
   - (ii) Supplementary Protection Certificate

Chapter 2

Copyrights

Introduction

The copyright law came into force after the discovery of printing machine some where in the middle of 17th Century. It was first enforced in England and later on was adopted by other Countries of the World to control unauthorized use of book and other such matters by other person. Copyright refers to the legal right granted to an author, composer, playwright, publisher, or distributor to exclusive publication, production, sale, or distribution of a literary, musical, dramatic, or artistic work is called copyright.

Main Features

The main features of copyright laws in relation to validity, registration,transfer, infringement and control are briefly presented below:

1. Period

It provides protection for a specific duration. The duration of copyright is 70 years after the death of the Author. For a work for hire or for an anonymous or pseudonymous work, the copyright lasts for 95 years after publication or 120 years after creation whichever is earlier.

2. Registration

The registration is not compulsory for copyright. In other words, it is an unregistered right. Since there is no registration for copyright, there is no need of filling application and there is no fee for copyright. However, it is advisable to register the copyright with the copyright office. It will provide additional legal rights to the copyright holder.

3. Validity

The copyright is valid in all the Countries of the World.

4. Transfer

The copyright holder has the rights to authorize others to use the protected work. In other words, the author may transfer the copyright to any other party if he so desires.

5. Items Covered

It is applicable to books, movies, music, paintings, photographs, sound recordings and software.

6. Control

The copyright is held by the author or creator of the new work. It is inherited as ordinary property. The information cannot be reproduced as such without written permission of the author or creator. However, the information or idea can be used by any one for creating the new work. The owner of a copyright can exclude others from reproducing, adapting, publicly distributing, publicly displaying or performing the copyrighted work.

7. Enforcement

The copyright comes into effect immediately after release of the matter covered by copyright such as book, movie, sound recording, etc.

8. Rights Protected

The purpose of copyright law is to encourage creative work by granting a temporary monopoly in an author's original creation. It controls the right of:

(i) Reproduction of work *i.e.* copying
(ii) Creation of derivative work
(iii) Distribution of work
(iv) Performance of the work
(v) Display of the work, and
(vi) Digital transmission of performance.

The law of copyright protects the first two rights in both private and public contexts, whereas an author can only restrict the last four rights in the public sphere.

9. Infringement

Infringement refers to an unauthorized use of material protected by copyright, patent or trademark law. Infringement of the copyrighted material is possible. Because it an unregistered right. The official action cannot be taken against infringement of unregistered copyrights. The following acts by a person other than the author amount to infringement.

(i) Reproducing or copying of the work
(ii) Creating a new work derived from the original work such translating the work into a new language and transferring the work into a new medium of expression.

(iii) Sell or give away the work or a copy of the work. Before transfer of copyright to that person.

(iv) Performing or displaying the work in public without permission from the copyright owner.

Defenses to Infringement

The primary defense to copyright infringement is the fair use. The fair use is described as the privilege to use the copyrighted material in a reasonable manner without the consent of the owner. However use should not have adverse on the marketing of the original author. The following activities are considered as fair uses.

(i) Distributing copies of a section of an article in class for educational purposes,

(ii) Providing a quotation in a book review,and

(iii) Imitating a work for the purpose of parody or social commentary.

Requirements

There are two basic requirements for protection of work under copyright *viz.*, (1) originality and (2) fixed expression. These are explained as follows:

(1) Originality

The first requirement is originality. Originality means that the work should be independently creared by the author.For originality the work need not necessarily have novelty,artistic merit,truth or lawful contents. For example a replica of painting in the public domain may not be novel, but is copyrightable. Similarly, a false biography is copyrightable.

(2) Fixed Medium of Expression

The work should be in fixed medium of expression. For example if someone performs in a local park it is not coyrightable. But if the performance is recorded in a film, then the film is copyrightable. Single words, short phrases and titles are not protected under copyright, because these are not considered "original works of authorship"under the copyright Act.

Types of Work Covered

A wide variety of work is protected under copyright law. For protection under copyright the work should have originality and wide utility.The list of items protected under copyright law is presented below:

1. **Literary Work**: It includes books [novels, instruction manuals etc], computer programs, lyrics for songs, articles in news papers and some types of database.
2. **Dramatic Work**: It includes works of dance or mime.
3. **Artistic Work**: It includes graphics and pictures such as paintings, engravings, photographs, sculptures, collages, works of architecture,

technical drawings, diagrams, maps, logos etc.

4. **Sound Recordings**: It includes recordings on tape and compact disc. It also includes recordings of other copyright work such as musical or literary.
5. **Musical and choreographic work.**
6. **Audiovisual Work**: Such as Movies andvideos; and
7. **Broad casts.**
8. **Work for Hire**: A work which is created by an employee in the regular course of his employment, it is considered a **work for hire** and the employer, not the employee, is considered the author of the work for copyright purposes.

Copyrights and Patents

There are several differences between copyrights and patents which have been presented in Table 2.1.

Table 2.1: Comparison of Copyrights and Patents

Sl.No.	*Particulars*	*Copyrights*	*Patents*
1.	Requirement	Originality and wide utility	It is valid in all countries
2.	Period	70-120 years	15-20 years
3.	Registration	Not compulsory	Compulsory
4.	Validity	All countries	In original country
5.	Fee	No Fee	Fees is charged
6.	Legal rights	No legal Rights	There are legal Rights
7.	Transfer	Possible	Admisible
8.	Items covered	Literary, music, movie etc	Products/process/idea
9.	Enforcement	Immediately after release of work	After Registration
10.	Control	By Author/creator	By inventor
11.	Infringement	Very common	Possible

Copyrights and Trademarks

There are several differences between copyrights and trademarks which have been presented in Table 2.2.

Advantages of Copyright

1. Registration is not compulsory.
2. It protects original work.
3. There is no fee for copyright protection.
4. The protection is for a very long period.
5. It is valid in all countries.

Table 2.2: Comparison of Copyrights and Trademarks

Sl.No.	*Particulars*	*Copyrights*	*Trademarks*
1.	Requirement	Originality and wide utility	New and distinctive
2.	Period	70-120 years	Renewable indefinitely
3.	Registration	Not compulsory	Is compulsory
4.	Validity	All countries	All Countries
5.	Fee	No Fee	Fee is charged
6.	Legal rights	No legal Rights	Legal rights are there
7.	Transfer	Possible	Possible
8.	Items covered	Literary, music, movie etc	Names, images and short phrases
9.	Enforcement	After release of work	After registration
10.	Control	By Author/creator	By owner
11.	Infringement	Very common	Possible

Disadvantages of Copyright

1. Copyrights do not protect ideas, procedures, processes, systems, methods of operation, concepts, principles or discoveries.
2. It does provide legal rights without registration.
3. It protects only against copying.
4. Do not protect historical facts.

Summary

The copyright law came into force after the discovery of printing machine some where in the middle of 17th Century. It was first enforced in England and later on was adopted by other Countries of the World to control unauthorized use of book and other such matters by other person.

Copyright refers to a document that grants exclusive right to the author/creator to publish and sell literary or musical or artistic work.

The main features of copyright laws in relation to period, validity, registration,transfer, enforcement, fees, requirements, items covered, legal rights infringement and control have been discussed. A comparison of copyrights and patents and copyrights and trademarks has been presented in tabular form for easy grasp of the subject. Types of work protected and not protected under copyright law have been explained.

Copyright infringement and defenses for infringement have been presentedn Advantages and disadvantages of copyrights have been listed.

Glossary

Copyright: The exclusive rights granted to author/creator to reproduce, sell and distribute a work, prepare derivative works and display the work publicly.

Infringement: An unauthorized use of material protected by copyright, patent or trademark law.

Fair use: The privilege to use the copyrighted material in a reasonable manner without the consent of the owner.

Hire Work: A work which is created by an employee in the regular course of his employment,

Plagiarism: Copying another's words or ideas without attributing them to the true author.

Questions

1. **What is copyright law? Discuss briefly ther important feature.**
2. **Describe briefly the advantages and disadvantages of copyright law.**
3. **Describe copyright law on the basis of following points.**
    a. Period of protection
    b. B. Legal Implications
    c. Work to be protected
    d. Registration
4. **What are basic requirements for protection under copyrights?**
5. **Give differences between copyright and patents in tabular form.**
6. **Explain briefly the differences between copyrights and trademarks.**
7. **Write short notes on the following:**
    a. Copyright
    b. Infringement
    c. Fair use of copyrights
    d. Work of hire
    e. Plagiarism
8. **Define Infringement. Describe infringement of copyrights with examples.**
9. **Describe briefly main features of copyright law.**
10. **Explain the following:**
    a. Works protected under copyrights
    b. Work not protected under copyrights
11. **What do you mean by fair use of copyrights? Discuss with examples.**

Chapter 3

Patent Rights

Introduction

The first Patent Law was adopted in Republic of Venice in 1474 and later on patent law was introduced in other countries. The main purpose of patent law is to promote the progress of science and useful arts by providing exclusive rights to inventors for their inventions for a limited period so that inventors can get advantage of their inventions. Patent is a legal document issued by the patent office that grants exclusive rights for the production, sale and profit from the invention of a product or process for a specific period of time. Patents also grant the right to prevent others from copying the invention.

Types of Patents

Patents have been classified in various ways. On the basis of duration, patents are of two types, *viz.* (1) full-term patents and (2) short-term patents. These are briefly discussed below:

Full-Term Patents

The maximum period for grant of a patent is 20 years. Those patents that are granted for a period of 20 years are referred to as full-term patents. For the grant of full-term patent, the applicant must provide evidence of the novelty of invention. Thus full-term patents have two main features as follows:

(i) They are granted for full-term of 20 years.

(ii) Evidence of the novelty of invention is required.

Short-Term Patents

Those patents that are granted for less than 20 years are called short-term patents. Such patents are granted for smaller inventors There are two main features of such patents as given below:

(i) They are granted maximum for a period of 10 years.
(ii) Evidence of the novelty of invention is not required.

The patents are also classified as (i) utility patents, (ii) design patents, and (iii) plant patents. These are briefly discussed as follows:

Utility Patent

A type of patent that is granted to inventions that perform useful functions is referred to as utility patent. This is also known as patent for invention. Main feature of utility patent are listed below:

(i) Such patents are issued for the invention of new and useful process, machine, manufacture or composition of matter and their improvement.
(ii) It is granted for full-term of 20 years.
(iii) About 90 per cent of the patents are utility patents.

Design Patent

A type of patent that is granted to visual ornamental characteristics such as shape and surface of an article of manufacture is called design patent. It is also called as picture patent. A design for surface ornamentation is inseparable from the article to which it is applied and cannot exist alone. It is definite pattern of surface ornamentation, applied to an article of manufacture. Main features of utility patent are listed below:

(i) This is granted for a period of 14 years.
(ii) It is also known as picture patent.
(iii) Design patents do not protect an idea or an invention. They only protect ornamental design of exactly what is pictured.
(iv) They are very weak and appropriate in limited circumstances.

Plant Patent

A type of patent that is granted to new or newly developed, distinct plants reproducing asexually. Main features of plant patents are listed below:

(i) These patents are of recent origin.
(ii) They are useful in protecting novel plants or new plants.
(iii) Protected for a period of 5 years.

Main Features

A patent consists of an abstract, a description of the invention, disclosure of prior art, drawings, and one or more claims. The main features of patent laws in relation to validity, registration,transfer, infringement and control are briefly presented below;

1. **Period:** It provides protection for a specific duration. The period of protection varies from 15 to 20 years. In India it is 7-14 years. The long

term patents are protected for 20 years in many countries and short term patents for 14 years.

2. **Registration:** The registration is compulsory for the patent. In other words, patent is a registered right. An application is submitted in prescribed form to the patent office for the registration of the paten and some fee is also paid for registration.
3. **Validity:** The patent is valid in the country where it has been granted. In future the patent granted by WTO will be valid in all member countries.
4. **Transfer:** The inventor has the power to transfer the right of patent to any other person if he so desires. In other words, the inventor may transfer ownership to any one [including a corporation]. Employee often assign the rights of their invention to their employers as part of their employment contracts.
5. **Items Covered:** Patent is applicable to new process, products, invention, device and innovations which have commercial or industrial applications.
6. **Control:** The inventor of the process, product or invention has control over the patent Patents provide exclusive rights to the inventor for the production, sale and profit from his invention or a product or process for a specific period of time.
7. **Enforcement:** The patent becomes effective immediately after its registration.
8. **Infringement:** Infringement of the patent refers to an unauthorized use of the process or product by other person. A patent owner may recover money damages in the form of a reasonable royalty. If the infringement is found willful, three times money is recovered.
9. **Rights Protected:** A utility patent once granted, provide the owner of the patent the right to prevent others from making, using, selling and importing a product covered by the utility patent.

Requirements for Patents

Four criteria, *viz.*, patentable subject matter, novelty, usefulness and non-obviousness are sufficient for granting patent.

Patentable Subject Matter

1. Patentable Matters

The patentable subject matter include five main items, *viz.* (1) processes, (2) machines (3) articles of manufacture, (4) business methods and (5) software related ideas. Besides, new algorithms, data structure, new applications, business procedure and new technology can be patented. The same software may qualify for both copyright and patent protection, if it satisfies requirements for both of these.

Processes include business methods, most software, medical techniques, sports techniques and the like.machines include devices and apparatuses. Articles of manufacture include mechanical devices, electrical/electronic devices and composition of matter such as chemicals, medicines, DNA, RNA, etc.

2. Non-patentable Matters

The following matters cannot be patented.

(i) Ideas, mental processes, laws of nature and physical phenomena. For example one cannot a new mineral, insect or plant discovered or found in the wild.

(ii) Literary, dramatic, musical and artistic work. These are subject to copyright protection.

(iii) Inventions that are not useful or that are offensive to public morality.

Procedure of Filing Patent

An inventor cannot patent an inventions if he or she discloses it to the public more than one year before filing application for patent prosecution. Predisclosure is known as on sale which bar to patentability. The legal process of obtaining a patent from a patent office is called patent prosecution. An application for a patent is prepared by a professional patent agent or patent attorney and filed with a patent office where it is examined by a patent examiner for patentability.Patent protection in multiple countries requires filing of separate patent applications in each country. The patent prosecution process involves following steps:

1. Filing a duly filled up patent application by inventor or applicant.
2. Establishing of a prior art search report by the patent office.
3. Publication within 18 months from earliest claimed filing date.
4. Review by the examiner or the examining Division, including communication with applicant to modify the claim language, if needed.
5. Grant of the patent [if the patentability criteria are met] and publication of the issued patent.
6. Opposition period, during which anybody (*e.g.* other companies) can challenge the patent. In several countries, oppositions can be filed before the grant of the patent.

Publicly disclosed inventions including patented inventions are known as prior art. If prior art is not disclosed by the applicant in his patent application the patent can be invalidated for hiding information.

Criteria for Examination

The following criteria are taken into account in the examination process of the patent application submitted by the inventor or applicant.

1. Verifying that claims are for a patentable subject matter.
2. Ensuring unity of invention: there should separate application for each invention.

3. Ensuring that the drawings, description and claims meet all formal requirements.
4. Utility or industrial applicability.
5. Novelty [newness] and
6. Non-obviousness or inventive step.

Four criteria, *viz.*, patentable subject matter, novelty, non-obviousness and sufficient disclosure are sufficient fro granting patent.

Patent versus Trademarks

There are several differences between patent and trademarks which have been presented in Table 3.1 below.

Table 3.1: Comparison of Patents and Trademarks

Sl.No.	*Particulars*	*Patents*	*Trademarks*
1.	Requirement	It is valid in all countries	New and distinctive
2.	Period	15-20 years	Renewable indefinitely
3.	Registration	Compulsory	Is compulsory
4.	Validity	In original country	All Countries
5.	Fee	Fees is charged	Fee is charged
6.	Legal rights	There are legal Rights	Legal rights are there
7.	Transfer	Admissible	Possible
8.	Items covered	Products/process/invention/innovation	Names, images and short phrases
9.	Enforcement	After Registration	After registration
10.	Control	By inventor	By owner
11.	Infringement	Possible	Possible

Advantages of Patents

1. It excludes competitors.
2. It builds licensing portfolio.
3. It enhances product value.

Disadvantages of Patents

1. It involves cost of prosecution and renewal.
2. It requires public disclosure of the invention.
3. The validity is for a short duration [15-20 years].
4. Patents are territorial in nature.

Summary

The first Patent Law was adopted in Republic o Venice in 1474 and later on introduced in other countries. The main purpose of patent law is to promote the progress of science and useful arts by providing exclusive rights to inventors for their inventions for a limited period so that inventors can get advantage of their inventions.

Patent is a document that allows the patent owner to prevent others from making, using or selling the invention protected by the patent. Patents have been classified in various ways. On the basis of duration, patents are of two types, *viz.* (1) full-term patents and (2) short-term patents. The patents are also classified as (i) utility patents, (ii) design patents, and (iii) plant patents.

The main features of patents have been discussed in relation to period, validity, registration,transfer, enforcement, fees, requirements, items covered, legal rights infringement and control. A comparison of patent and trademarks has been presented in tabular form for easy grasp of the subject. List of items registered under patent has been given. The basic requirements and procedure of paten registration have been outlined. Advantages and disadvantages of copyrights have been presented.

Glossary

Business Method Patent: A specific type of software patent on a computer way of transacting business.

Design Patent: A type of patent that is granted to visual ornamental characteristics such as shape and surface of an article of manufacture.

Full-term Patent: Patents that last for a period of 20 years.

License: A promise by the patent owner (the licensor) not to sue the licensee for exercising on of the patent owner's righrs.

Patent infringement: An un authorized use of patented invention/innovations.

Patent Prosecution: The legal process of obtaining a patent from a patent office.

Patent: A patent is a set of exclusive rights granted by a state to a patentee (the inventor or assignee) for a fixed period of time.

Plant Patent: A type of patent that is granted to newly developed plants having novel traits and fulfilling certain requirements.

Pre- disclosure: Disclosure of an invention before patenting.; also known as on sale.

Prior Art: Publicly disclosed inventions, including patented inventions. It can be cited against a new patent application.

Provisional Patent Application: An application filed before filing regular patent application to establish a filing date which may later be claimed as priority date.

Short Term Patent: Patents that last for a maximum period of 10 years.

Utility Patent: A type of patent that is granted to inventions that perform useful functions.

Questions

1. **Define patent and describe its main features.?**
2. **Describe briefly the advantages and disadvantages of patent rights.**
3. **Describe patent rights on the basis of following points.**
   a. Period of protection
   b. Legal Implications
   c. Work to be protected
   d. Registration
4. **What are basic requirements for registration of a patent?**
5. **Describe briefly the procedure of filing the patent.**
6. **Explain briefly the differences between copyrights and trademarks.**
7. **Write short notes on the following:**
   a. Patent
   b. Patent Infringement
   c. Types of patent
   d. Duration of patent
8. **What is patent infringement ? Explain the same with examples.**
9. **Give a brief account of different types of patents.**
10. **Explain the following:**
    a. Patentable Items
    b. Non patentable Items
11. **Define the following terms:**
    a. Prosecution
    b. Prior Art
    c. Utility patent
    d. Design patent
    e. Plant patent
    f. Full-term patent
    g. Short term patent

Chapter 4

Trademark Rights

Introduction

The first trademark was registered in Britain in 1876 for the Bass Red Triangle. Now trademarks are registered almost in every country. A trademark is a distinctive sign of some kind which is used by an individual, business organization or other legal entity to uniquely identify the source of its products and/or services to consumers, and to distinguish its products or services from those of other entities.

Main Features

The main features of trademarks rights in relation to validity, registration,transfer, infringement and control are briefly presented below:

1. **Period:** Initial term for a trademark registration is 10 years. It can last as long as the owner wants. Between the fifth and sixth year after the date of initial registration, the trademark registrant will need to file documents to keep the mark alive. A renewal then be filed every 10 years. Thus it can last indefinitely.
2. **Registration:** It is not essential to register a trademark. However, it is advisable to register the trademark to get legal rights against infringement of trademarks. Almost all countries in the world register and protect trademarks. The prescribed Fee has to be paid for registration of trademark.
3. **Validity:** The trademark rights are National, not International. Therefore, registration in one country does no confers rights in other countries. A separate application will be needed for registration in other country.
4. **Transfer:** The owner of the trademark has right to authorize another person to use it on payment (royalty) basis.
5. **Matters Covered:** A trademark includes a distinctive word, phrase, logo, domain name, graphic symbol, slogan, image or other device used to

identify the source of a product from others. Thus a variety of items are covered by trademark. Recently colors [such as John deere green], sounds and scents have also been registered as trademarks.

6. **Control:** A trademark owner has rights to prevent unauthorized use of trademark.
7. **Enforcement:** The trademark becomes effective when it is first publicly put to continuous commercial use, even without registration.
8. **Infringement:** An unauthorized use of a trademark by a person other than the trademark owner is called infringement of trademark. Infringement occurs when someone uses the same or similar trademark for a similar good or service. Infringement does not require the infringer to be competitor, but infringement does require a similar product.
9. **Rights Protected:** Protection of trademarks provides exclusive right to the owner of the trademark to use it to identify goods or services.
10. **Requirements:** Trademarks apply to products and services. There are some requirements for registration of a trademark. It must be distinctive, not descriptive and there must be no earlier rights or multiple appearance. Moreover, it should not confuse with other trademark. It should not be prohibitive (reserved) and generic.

Types of Trademarks

There are many types of trademarks such as, (1) service mark, (2) trade dress, (3) collective mark and (4) certification mark. These are defined as follows:

Service Mark

A trademark that is used in relation to services rather than products is called a service mark. It promotes services or events. Some examples of service marks are as follows:

(i) Amazon.com [retail website],
(ii) B.Jack in the Box [fast food service],
(iii) Kinko's [photocopying service]
(iv) Blockbuster [video rental service] and
(v) Fed Ex logo [delivery service]

Trade Dress

A trademark that is known by its distinctive packaging is referred to as trade dress, for example the yellow packaging of Kodak film. The trade dress often serves the same function as a trademark or service mark.

Collective Mark

A trademark used by members of a group or organization is known as collective mark. It may be a symbol, label, phrase or other mark. These are used to identify goods, products or services rendered by a group. Collective marks are often used to show membership in a union, association or other organization.

Certification Mark

A certification mark is a symbol, name or device used by an organization to guarantee for the quality products and services provided by others such as Good Housekeeping Seal of Approval.

Based on the relationship, between the mark and the underlying product, trademarks are of four types, *viz.* (1) arbitrary, (2) suggestive, (3) descriptive, and (4) generic. These are defined as under.

Arbitrary Mark

An arbitrary mark is one that bears no logical relationship to the underlying product. For example, the words Ëxxon"and "Kodak"bear no inherent connection to their underlying products. It is also known as fanciful mark.

Suggestive Mark

This is a mark that evokes or suggests a characteristic of the underlying good. For example the word "Copper tone" suggests the colour of a deep sun tan.

Descriptive Mark

A descriptive mark directly describes rather than suggests, a characteristic or quality of underlying product such as its colour, odor, function, diamensions or ingradients. For example "Holiday Inn"and A"ll Bran"describe hotel rooms and breakfast cereal.

Generic Mark

A generic mark is a mark that describes the general category to which the underlying product belongs For example, the term "Computer" is a generic term for computer equipment. Such terms are not entitled for protection.

Procedure of Filing Trademarks

The application for registration of trademark is to be submitted in the Patent and Trademark Office. The prior use of the trademark for commercial use is required for registration. The registration process consists of the following steps:

1. Filling of Application

The prescribed application form is used and the application has to be carefully completed and signed. The following documents should be attached with the application. Attached to the goods, containers for the goods.

(i) Drawing of the mark on sepsrate piece of paper.

(ii) **Prescribed Fee**: The Fee is not refundable even if the application is rejected.

(iii) **Specimens**: The manner in which the mark is to be used should be indicated. The application must clear reproduction of the sign filed for registration including, any colors, forms or three. dimensional features.

The application must also contain a list of goods or services to which the sign would apply. The sign must be distinctive to identify the product from other products. Each application must include three specimens.

2. Examination of the Application

The application is examined in the PTO by an examining attorney. If the application is found correct, the attorney will approve the mark for publication in the official Gazette. A trademark search may also be conducted to know whether the mark is already in the use by some other party.

3. Opposition

If any party thinks that the mark will damage it rights, opposition can be filed within 30 days from the date of publication in the Gazette.

4. Registration

If no opposition is filed, the PTO will register the trademark and issue a registration certification a reasonable time.

Trademarks and Trade Secret

There are several differences between trademarks and trade secret which have been presented in Table 4.1 below:

Table 4.1: Comparison of Trademarks and Trade Secret

Sl.No.	*Particulars*	*Trademarks*	*Trade Secret*
1.	Requirement	New and distinctive	Any thing confidential of commercial use
2.	Period	Renewable indefinitely	Indefinite duration
3.	Registration	Is compulsory	Not required
4.	Validity	In original country	With the company concerned
5.	Fee	Fee is charged	No Fee
6.	Legal rights	Legal rights are there	No legal rights
7.	Transfer	Possible	Possible
8.	Items covered	Names, images and short phrases	Formula, pattern, device, business methods, source of supply, customers list etc.
9.	Enforcement	With commercial use	With commercial use
10.	Control	By owner	By owner
11.	Infringement	Possible	Not possible

Advantages of Trademarks

1. Excludes competitors.
2. Renewable indefinitely.

3. Can be protected by common law rights.
4. Enhances reputation of the company.

Disadvantages of Trademarks

1. It involves cost of prosecution and renewal.
2. Infringement is possible.
3. Registered in the country where product is used.

Summary

The first trademark was registered in Britain in 1876 for the Bass Red Triangle. Now Trademarks are registered almost in every country. A trademark is a distinctive sign of some kind which is used by an individual, business organization or other legal entity to uniquely identify the source of its products and/or services to consumers, and to distinguish its products or services from those of other entities.

The main features of trademarks rights in relation to validity, registration,transfer, infringement and control have been discussed. Procedure of registering trademarkshas been outlined.

There are many types of trademarks such as, (!) service mark, (2) trade dress, (3) collective mark and (4) certification mark. Based on the relationship, between the mark and the underlying product, trademarks are of four types,*viz.* (1) arbitrary, (2) suggestive, (3) descriptive, and (4) generic.

Comparison of trademarks and trade secret has been presented in tabular form for easy grasp of the readers. and advantages and disadvantages of trademarks have been described.

Glossary

Arbitrary Mark: An arbitrary mark is one that bears no logical relationship to the underlying product.

Certification Marks: A certification mark is a symbol, name or device used by an organization to guarantee for the quality products and services provided by others

Collective Mark: A trademark used by members of a group or organization is known as collective mark. It may be a symbol, label, phrase or other mark.

Descriptive Mark: A descriptive mark directly describes rather than suggests, a characteristic or quality of underlying product such as its color, odor, function, dimensions or ingradients.

Generic Mark: A generic mark is a mark that describes the general category to which the underlying product belongs.

Service Marks: A trademark that is used in relation to services rather than products.

Suggestive Mark: This is a mark that evokes or suggests a characteristic of the underlying goods.

Trade Dress: A trademark that is known by its distinctive packaging is referred to as trade dress,.for example the yellow packaging of Kodak film

Trade Search: A search that is conducted to know whether the mark is already in the use by some other party or not.

Trademarks: Trademark is used to identify its product and to distinguish them from others. It is the name of a product made by a particular person or company

Questions

1. **Define trademarks and describe their main features.?**
2. **Describe briefly the advantages and disadvantages of trademarks.**
3. **Describe trademarks on the basis of following points.**
   a. Period of protection
   b. Legal Implications
   c. Work to be protected
   d. Registration
4. **What are basic requirements for registration of trademark?**
5. **Describe briefly the procedure of registering trademarks.**
6. **Explain briefly the differences between trademarks and trade secret.**
7. **Write short notes on the following:**
   a. Arbitrary Mark
   b. Suggestive Mark
   c. Descriptive Mark
   d. Generic Mark
8. **What is trademark infringement ? Explain the same with examples.**
9. **Describe briefly main features of trademarks.**
10. **Explain the following:**
    a. Service Marks
    b. Certification Marks
11. **Give a brief account of different types of trademarks.**

Chapter 5

Industrial Design

Introduction

An industrial design refers to any original shape, picture, or some combination applied to a useful article of manufacture. In other words, it refers to the design of the mass-produced products of our every day environment, from sinks and furniture to computers. Industrial designs optimize the function, value, and appearance of products and systems for the mutual benefit of both user and manufacturer.Novelty and originality are important criteria in a design for registration. In addition, only those designs that are applied to an article by an industrial process are protected.

Main Features

Main features of industrial designs in relation to validity, registration,transfer, infringement and control are briefly presented below;

1. **Period:** Initially the design is registered for a term of 5 yearly basis. Later on it can be renewed on 5 years by paying the prescribed Fee. In some cases, it may be registered up to 15 years.
2. **Registration:** Registration is essential to get legal rights. There is no protection without registration. A nominal processing Fee is charged for registration.
3. **Validity:** The protection of the design is valid only in the country where it has been registered. The protection in other countries can be obtained by filling separate application in each country.
4. **Transfer:** A registered design is a property like any other commodity which can be purchased, sold, hired or licensed. It is advisable to register the sale of rights in the registration office before use of the design by new owner.

5. **Matters Covered:** Industrial designs include any original shape, picture, or some combination applied to a useful article of manufacture. Industrial designs do not cover following things.
   (i) The functional features of an article.
   (ii) A principle of construction, or how an article is built.
   (iii) The materials used in the construction of an article.
   (iv) Color per se, and Ideas.
6. **Control:** The owner has the legal rights to prevent others from unauthorized use of his design.
7. **Enforcement:** The industrial design comes into force immediately after it registration in the patents and trademarks office.
8. **Infringement:** Unauthorized use of a design is called infringement of such design. The owner has the right to take legal action against the infringer and claim damages. However, the legal action must be taken within three years of the alleged offence.
9. **Rights Protected:** It provides exclusive rights to the owner such as to make,imports, sell or hire out or allow others to use the design under terms settled with the registered owner.
10. **Requirements:** There are two requirements of industrial design for registration, *viz.* novelty and commercial application. In other words, it should different from previously registered designs and it should not be made public before registration. It is made open to public after registration.

Procedure for Registration

The registration is done in Patents and trademarks Office. The owner of the design has to complete the following formalities:

1. He has to submit duly filled up application in prescribed form in the Patents Office Prescribed Fee has to paid along with application.
2. The specimen of the design is to be provided. The only visual aspects of the design such as shape, configuration, pattern or ornament or any combination thereof should be described. It must state clearly what the design is. Too much details are not required.
3. The registry office will make a search about novelty of the design.
4. If requirements are met, a certificate of registration is issued otherwise registration will be refused which is appeal able.
5. Initially 5 years registration is granted that can be renewed on five yearly basis.
6. Normally, the registration process should be completed within six months. However, a total of 12 months [extendable to 15 months] time is allowed for completion of registration process. The design should not be made public before registration otherwise registration may later be proved invalid.

Industrial Design and Patent

There are several differences between industrial design and patent which have been presented in Table 5.1 below.

Table 5.1: Comparison of Industrial Design and Patents

Sl.No.	*Particulars*	*Industrial Design*	*Patents*
1.	Requirement	A design should be novel	It is valid in all countries
2.	Period	5-15 years	15-20 years
3.	Registration	Required	Compulsory
4.	Validity	In the country of registration	In original country
5.	Fee	Fee is charged	Fees is charged
6.	Legal rights	Owner has legal rights	There are legal Rights
7.	Transfer	Possible	Admissible
8.	Items covered	Feature of shape pattern, configuration or ornament on a finished article.	Products/process/invention// innovation
9.	Enforcement	After registration	After Registration
10.	Control	By right owner	By inventor
11.	Infringement	Possible	Possible

Industrial Designs and Copyrights

There are several differences between industrial design and copyrights which have been presented in Table 5.2 below.

Table 5.2: Comparison of Industrial Design and Copyright

Sl.No.	*Particulars*	*Industrial Design*	*Copyrights*
1.	Requirement	A design should be novel	Originality and wide utility
2.	Period	5-15 years	70-120 years
3.	Registration	Required	Not compulsory
4.	Validity	In the country of registration	All countries
5.	Fee	Fee is charged	No Fee
6.	Legal rights	Owner has legal rights	No legal Rights
7.	Transfer	Possible	Possible
8.	Items covered	Feature of shape, pattern, configuration or ornament on a finished article.	Literary, music, movie etc.
9.	Enforcement	After registration	Immediately after release of work
10.	Control	By right owner	By Author/creator
11.	Infringement	Possible	Very common

Industrial Designs and Trademarks

There are several differences between industrial design and trademarks swhich have been presented in Table 5.3 below:

Table 5.3: Comparison of Industrial Design and Trademarks

Sl.No.	*Particulars*	*Industrial Design*	*Trademarks*
1.	Requirement	A design should be novel	New and distinctive
2.	Period	5-15 years	Renewable indefinitely
3.	Registration	Required	Is compulsory
4.	Validity	In the country of registration	In original country
5.	Fee	Fee is charged	Fee is charged
6.	Legal rights	Owner has legal rights	Legal rights are there
7.	Transfer	Possible	Possible
8.	Items covered	Feature of shape, pattern, configuration or ornament on a finished article.	Names, images and short phrases
9.	Enforcement	After registration	With its commercial use
10.	Control	By right owner	By owner
11.	Infringement	Possible	Possible

Industrial Design and Trade Secret

There are several differences between industrial design and trade secret which have been presented in Table 5.4.

Table 5.4: Comparison of Industrial Design and Trade Secret

Sl.No.	*Particulars*	*Industrial Design*	*Trade Secret*
1.	Requirement	A design should be novel	Any confidential knowledge of commercial use
2.	Period	5-15 years	Indefinite duration
3.	Registration	Required	Not required
4.	Validity	In the country of registration	With the company concerned
5.	Fee	Fee is charged	No Fee
6.	Legal rights	Owner has legal rights	No legal rights
7.	Transfer	Possible	Possible
8.	Items covered	Feature of shape, pattern, configuration or ornament on a finished article.	Formula, pattern, device business methods, source of supply, customers list etc
9.	Enforcement	After registration	With commercial use
10.	Control	By right owner	By owner
11.	Infringement	Possible	Not possible

Advantages of Industrial Design

1. Enhance aesthetic value of the product
2. Provide legal rights to the owner against unauthorized use by others
3. Make an article attractive and increase it marketability
4. Can be purchased, sold or licensed
5. Enhance reputation of the company.
6. Encourages creativity in the industrial sector and in arts and crafts.

Disadvantages of Industrial Design

1. Require high technical skill
2. Are liable to infringement
3. Involve renewal cost
4. Are valid in the country of registration only
5. Lead to monopoly

Summary

An Industrial design is the features of shape, configuration, pattern or ornament [or any combination thereof] applied to a finished article. For example, it may the shape of a table, or the ornamentation on the handle of a spoon. The article may be made by hand, tool or machine.

Industrial designs include any original shape, picture, or some combination applied to a useful article of manufacture. Industrial designs do not cover things such as (i) the functional features of an article, (ii) a principle of construction, or how an article is built, (iii) the materials used in the construction of an article, (iv) color per se, and (v) ideas.

Main features of industrial designs in relation to validity, registration, transfer, infringement and control have been discussed. The procedure of registering Industrial design and advantages and disadvantages of industrial designs have been discussed.

Glossary

Industrial Design: The features of shape, configuration, pattern or ornament [or any combination thereof] applied to a finished article.

Industrial Design Protection: A form of monopoly like other types of intellectual property.

Protected Design: A design registered with Patents and Trademarks Office.

Questions

1. Define Industrial design and describe its main features.?

2. What are the advantages of industrial design?

3. **Discuss in brief disadvantages of industrial design.**
4. **Describe industrial design on the basis of following points.**
   a. Period of protection
   b. Legal Implications
   c. Work to be protected
   d. Infringement
5. **Describe the procedure of registering industrial design.**
6. **Explain briefly the differences between industrial design and copyright.**
7. **Compare industrial design and patent in tabular form.**
8. **Compare Industrial design and trademarks in tabular form.**
9. **Give in brief comparison of industrial design and trade secret.**
10. **Write short notes on the following:**
    a. Industrial design
    b. Duration of protection
    c. Advantages of industrial design
    d. Protected design
11. **What are requirements for registration of Industrial designs ?**

Chapter 6

Utility Models

Introduction

A utility model is an exclusive right granted for an invention, which allows the right holder to prevent others from commercially using the protected invention, without his authorization, for a limited period of time. The important points about utility model are listed below:

1. Utility models are also known as petty patents, petty inventions, petty innovations, utility innovations, minor patents and small patent.
2. The rights conferred by utility models are similar to those granted by patent laws but has a shorter term.
3. They are registered with National Patent Office.
4. Presently, utility models are protected in about 170 countries of the World.
5. In some countries, utility model protection can only be obtained for certain fields of technology and only for products but not for processes.

Main Features

Main features of Utility Model in relation to materials covered, countries where protected, requirements, registration, rights covered, authorization, duration, validity and infringement etc. are briefly presented below:

1. Materials Covered

Utility model, means any form, configuration or disposition of elements of some appliance, working tools and implements as articles of everyday use, electrical and electronic circuitry of other object or part thereof in so far as they are capable of contributing some benefit or new effect of saving in time, energy, or labour or improving the hygienic or socio-physiological working conditions by means of new configuration, arrangement or device or a combination thereof and are industrially applicable.

2. Countries where Protected

A utility model is an intellectual property right to protect inventions. This right is available in more than 50 countries. It is very similar to the patent, but usually has a shorter term (often 6 or 10 years) and less stringent patentability requirements.

3. Requirements for Registration

Most countries having utility model laws require that the invention be new. However, many patent or utility model offices do not conduct substantive examination and merely grant the utility model after checking that utility model applications comply with formalities. The protection for utility models shall be available for utility models which are new and have industrial applications. Each utility model application shall be subjected to substantive examination for compliance with substantive examination requirements. Each designated state prior to registration shall be accorded the opportunity to give an opinion relating to registration in its territory of the utility model.

4. Registration

For the purposes of novelty and priority disclosures, utility model applications shall not be refused on grounds that the invention has been exhibited at an officially recognized exhibition before the date of filing. This exception will only apply if the exhibition occurred not more than six months before the date of filing.

(i) Filing Application

Utility model applications may be prepared and filed at local patent offices in countries where utility model protection is available. The application for a utility model should contain (i) name of the applicant, (ii)description of the utility model, (iii) drawings of the model, (iv) abstract, (v) state for which utility model is to be protected, and (vi) prescribed fee.

Conversion of Application

Application refused by the office or application which have been refused consideration can on request be converted into national applications. Patent application can also be converted into a utility model application or vice versa on request and such a converted application can benefit from the filing date of the initial application.

(ii) Examination of the Application

The application is examined by the experts. There is a pre-registration examination for Utility Models. This is conducted in three stages. The first stage involves examination to determine whether the application meets the formalities requirements given under 4 above.

The Second Stage involves substantive examination. The substantive examination is carried out to determine whether the utility model is industrially applicable and secondly whether it is new. Novelty in this case shall be the local Novelty in a designated state. On completion of the second stage, the application is forwarded to designated states.

The Third Stage involves examination by a designated state for novelty and for conformity with the national laws. This examination or verification should be completed within six months from the date of notification.

Each designated state is required to carry an examination to ascertain whether the Utility Model is New in the territory of the state designated.

(iii) Registration and Publication

Registration and Publication entails recording of the Utility Model in the Journal and the Register. It also entails issuance of a Certificate and Publication of a copy of the specification.

5. Registration Fee

The registration fee and maintenance charges are much lower than those for patents.

6. Rights Covered

Under the Utility Model Act, "the owner of a utility model shall have an exclusive right to commercially work the registered model" and is allowed to demand an injunction, compensation for damage, and compensation for unfair profits gained by the infringer.

7. Authorization

The inventor has the right to authorize another person to make use of his invention on royalty basis.

8. Duration

The utility model is registered for a shorter period than patent. The period of protection generally varies from 6 to 10 years in different countries.

9. Validity

The utility model will be valid in the country where it has been registered Provided The applicable national laws shall also apply.

10. Infringement

Infringement of utility model is possible. The owner or the inventor has legal rights to sue the infringer and recover damaged caused by way of infringement.

Patent and Utility Model

There several differences between patents and utility models. The main differences between them are as follows:

1. The requirements for acquirng a utility model are less strict than for patents. Novelty and non-obviousness" may be much lower or absent altogether. In practice, protection for utility models is often sought for innovations of a rather incremental character which may not meet the patentability criteria.

2. The term of protection for utility models is shorter than for patents and varies from country to country (usually between 6 and 10 years without the possibility of extension or renewal).
3. In most countries where utility model protection is available, patent offices do not examine applications as to substance prior to registration. This means that the registration process is often significantly simpler and faster, taking, on average, six months.

Table 6.1: Comparison of Utility Models and Patent

Sl.No.	*Particulars*	*Utility Model*	*Patents*
1.	Requirement	Novelty	Novelty
2.	Period	6-10 years	15-20 years
3.	Registration	Is compulsory but easy	Compulsory and difficult
4.	Validity	In country of registration	In original country
5.	Fee	Less Fee is charged	More Fees is charged
6.	Legal rights	Legal rights are there	Legal Rights are there
7.	Authorization	Admissible	Admissible
8.	Items covered	Utility inventions	Products/process/invention/ innovation
9.	Protection of plants and animals	Not permitted	Permitted
10.	Enforcement	After registration	After Registration
11.	Control	By Inventor	By inventor
12.	Infringement	Possible	Possible

4. Utility models are much cheaper to obtain and to maintain.
5. The Utility Model Act extends protection only to "devices relating to the shape or construction of articles or combination of articles." Any methods such as a manufacturing process are not covered by the Utility Model Act.
6. In the patent system, a substantive examination principle has been introduced. In the utility model system, on the other hand, a non-substantive examination principle has been introduced.
7. In addition to above, patent and utility models differs differ from each other in the durations of rights and administration fees.

Advantages of Utility Models

1. It includes simple devices.
2. Registration procedure is simple.
3. Registration is faster.
4. Much cheaper to obtain and maintain.

Disadvantages of Utility Models

1. It protects for shorter duration [6-10 years].
2. Generally it cannot be renewed.
3. Plants and animals cannot be registered.
4. Processes are not protected.

Summary

A utility model is an exclusive right granted for an invention, which allows the right holder to prevent others from commercially using the protected invention, without his authorization, for a limited period of time. In its basic definition, which may vary from one country (where such protection is available) to another, a utility model is similar to a patent. In fact, utility models are also referred to as "petty patents" or "innovation patents, minor patent and small patent."

Main features of Utility Model in relation to materials covered, countries where protected, requirements, registration, rights covered, authorization, duration, validity and infringement have been explained.

A comparison of utility models and patents has been presented in tabular form. Advantages and disadvantages of utility models have been outlined.

Glossary

Utility Model: A utility model is an exclusive right granted for an invention, which allows the right holder to prevent others from commercially using the protected invention, without his authorization, for a limited period of time.

Utility Model Infringement: Unauthorized use of a utility model by any person or company.

Questions

1. **Define Utility Model and describe their advantages.**
2. **Describe briefly main features of Utility Models.**
3. **Explain briefly the procedure of registration of Utility Models.**
4. **Describe Utility Models on the basis of following:**
   (i) Registration (ii) Duration of protection
   (iii) Rights covered (iv) Authorization
5. **Explain Utility Models on the basis of following:**
   (i) Infringement (ii) Registration fee
   (iii) Validity (iv) Legal Rights
6. **Describe briefly the requirements for registration of Utility Model.**
7. **Give a brief comparison of Utility Models and Patents in a tabular form.**

8. **Discuss briefly advantages and disadvantages of Utility Models.**

9. **Describe briefly important points about Utility Models.**

10. **Write short notes on the following:**

    (i) Utility Model (ii) Infringement

    (iii) Patent (Intellectual Property

Chapter 7

Geographical Indication

Introduction

Geographical Indication may be defined as the name of a region, a specific place or, in exceptional cases a country, used to describe an agricultural, natural or manufactured goods, product or a food stuff. In other words, a geographical indication is a sign used on goods that have a specific geographical origin and often possess qualities or a reputation that are due to that place of origin.

The World Intellectual Property Organization (WIPO) helps in protection of Geographical Indications [GI]. About 170 geographical indications have been registered. Now this number would be much more.

In India, the Parliament has passed the Geographical Indications of Goods (Registration and Protection) Act, 1999 [in December 1999]. This Act seeks to provide for the registration for better protection of geographical indications relating to goods in India. the Act would be administered by the Controller General of Patents, Design and Trade Marks who is the Registrar of Geographical Indications. The Geographical Indications Registry is located at Chennai.

So far geographical indications have been registered in some advanced countries because they have developed the system of protecting GI. In the past, wines, spirits, cheeses, tobacco, which account for 88 per cent, have been registered as GI. Wines and spirits account for almost 71 per cent of all registrations.

In addition to wines and spirits, there is need for protection of agricultural products as GI. For this there is need to establish a multilateral system of identification and registration of GI. Geographical Indication may be an agricultural, natural or manufactured goods or product.

Main Features

Main features of geographical indication in relation to validity, registration, transfer, infringement and control are briefly presented below:

1. **Period:** The registration of a geographical indication is valid for a period of 10 years. It can be renewed from time to time for further period of 10 years each. If a registered geographical indication is not renewed it liable to be removed from the register.
2. **Registration:** The registration is not compulsory. The registration provides better legal protection to take action against infringement. In other words, it confers legal protection to geographical indications. And prevents unauthorized use of a registered geographical indication by others.
3. **Validity:** The registration of the geographical indication is valid in the country of its origin.
4. **Transfer:** A geographical indication cannot be sold, purchased, transferred, licensed, pledged, mortgaged or put to such other agreement.
5. **Matters Covered:** It is used to identify agricultural, natural and manufactured goods.
6. **Control:** An authorized user has the exclusive rights o use the geographical indication in relation to goods for which it is registered.
7. **Enforcement:** The protection of geographical indication becomes effective after its registration.
8. **Infringement:** Infringement of a geographical indication is possible. The registered proprietor and authorized users can initiate action against infringement
9. **Rights Protected:** The proprietor or authorized user has the rights to use the geographical indication and prevent other from using such material without permission.
10. **Requirements:** It originates from a definite geographical territory and should have a special quality or reputation or other characteristics. The main features are as given below:
    (i) It is an indication.
    (ii) It originates from a definite geographical territory.
    (iii) It is used to identify an agricultural, natural or manufactured goods/product.
    (iv) The manufactured goods should be processed or processed in that territory.
    (v) It should have a special quality or reputation or other characteristics.

Procedure for Registration

The registration is done by the registrar of Geographical Indications. For registration any association of persons, producers, organization or authority established by or under the law can apply The applicant must represent the interest of the producers. The registration consists of following steps:

1. **Filing Application:** The application in the prescribed form is filed to the registrar of Geographical Indications along with prescribed Fee and other details.

2. **Examination:** The application is examined by the registrar office by conducting search. The application is accepted, if it fulfils the requirements. Then it is advertised in the GI journal for objection if any.
3. **Acceptance:** If there is no objection, the application is accepted for grant of registration.
4. **Registration:** Then the GI is entered in the Register. The register of geographical indication is divided into two parts, *viz.* part A and part B. Part A consists of particulars relating to registered geographical indication and part B consists of particulars of the registered authorized users.
5. **Issue of Certificate:** Finally a certificate is issued to the registered user. The registration of GI is for a period of 10 years which is renewable for further periods of 10 years each.

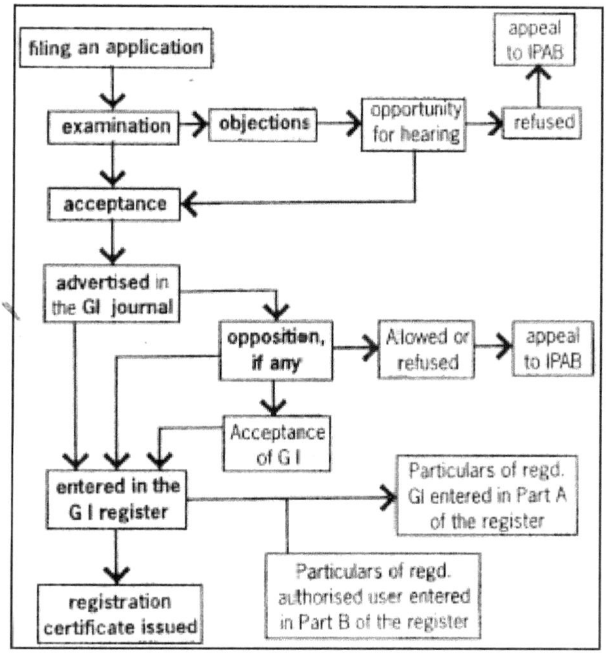

Figure 7.1: Registration Procedure for Geographical Indication.

Geographical Indication and Patents

There are several differences between geographical indication and patents which have been presented in Table 7.1.

Geographical Indication and Trademarks

There are several differences between geographical indication and trademarks as given below:

1. **Number:** The total number of registered trademarks is more than 16 millions. However, only about 6 millions trademarks are currently in use

Table 7.1: Comparison Geographical Indication and Patents

Sl.No.	Particulars	Geographical Indication	Patents
1.	Requirement	Reputed product of a specific place or territory	It is valid in all countries
2.	Period	10 years but renewable	15-20 years
3.	Registration	Not compulsory	Compulsory
4.	Validity	In the country of registration	In original country
5.	Fee	Fee is charged if registered	Fees is charged
6.	Legal rights	After registration only	There are legal Rights
7.	Transfer	Not allowed	Admissible
8.	Items covered	Agricultural, natural and manufactured goods	Products/process/invention/innovation
9.	Enforcement	After registration	After Registration
10.	Control	By authorized user or proprietor	By inventor
11.	Infringement	Possible	Possible

worldwide. Whereas about 170 GI have been registered so far. Thus only limited number of trademarks are useful

2. **Beneficiaries.** In case of GI, only local people or people of a region are beneficiaries, whereas in case of trademarks large number of people are beneficiaries.
3. **Role of State**. The State Government plays important role in protection of. Trademarks are promoted by individual company.
4. **Production**. The production of GI is collective by producers of a place or region, whereas the production of trademarks is by individual company.
5. **Protection.** The protection of GI is difficult as compared to trademarks.
6. **Creation**. The creation of new trademarks is easier than creation of new GI.

Besides these, there are several other differences between geographical indications and trademarks which are presented in Table 7.2.

Advantages of GI

1. GI protection provides legal rights to the users.
2. It prevents unauthorized use of registered GI.
3. It boosts exports.
4. It promotes economic prosperity of producers of goods in a geographical territory.
5. Geographical indications bring name to place of their origin or territory

Table 7.2: Comparison of Geographical Indication and Trademarks

Sl.No.	*Particulars*	*Geographical Indication*	*Trademarks*
1.	Requirement	Reputed product of a specific place or territory	New and distinctive
2.	Period	10 years but renewable	Renewable indefinitely
3.	Registration	Not compulsory	Is compulsory
4.	Validity	In the country of registration	In original country
5.	Fee	Fee is charged if registered	Fee is charged
6.	Legal rights	After registration only	Legal rights are there
7.	Transfer	Not allowed	Possible
8.	Items covered	Agricultural, natural and manufactured goods	Names, images and short phrases
9.	Enforcement	After registration	With its commercial use
10.	Control	By authorized user or proprietor	By owner
11.	Infringement	Possible	Possible

Disadvantages of GI

1. Prevents others for use.
2. It leads to monopoly.
3. Users are limited to place of origin.

Examples of Indian GI

In India some geographical indications have been identified which are listed below:

(i) Basmati Rice
(ii) Darjeeling Tea
(iii) Kanchipuram Silk Saree
(iv) Alphanso Mango
(v) Nagpur Orange
(vi) Kolhapuri Chappal
(vii) Bikeneri Bhujia
(viii) Agra Petha
(ix) Banarasi Saree

The above geographical indications can be classified into three groups, *viz.* (i) agricultural, (ii) natural,and (iii) muanufactured.

1. **Agricultural GI:** This group includes Basmati Rice, Nagpur Orange and Alphanso mango.
2. **Natural GI:** This group includes Darjeeling Tea.

3. **Manufactured GI:** This group includes Kanchipuram Silk Saree, Banarasi Silk Saree, Kolhapuri Chappal, Bikaneri Bhujia and Agra Petha.

Summary

The World Intellectual Property Organization (WIPO) helps in protection of Geographical Indications [GI]. About 170 geographical indications have been registered by Lisbon Agreement members up to 1997. Now this number would much more.

A geographical indication is a sign used on goods that have a specific geographical origin and often possess qualities or a reputation that are due to that place of origin. In other words, a geographical indication is a name or sign used on certain products or which corresponds to a specific geographical location or origin (*e.g.* a town, region, or country).

Main features of geographical indication in relation to validity, registration, transfer, infringement, control,etc. have been discussed. The procedure of registration of geographical indication has been explained. The comparison of geographical indication and patent and geographical indication and trademarks has been presented in tabular form. Advantages and disadvantages of geographical indications has been given. Examples of Indian geographical indications have been given.

Glossary

Geographical Indication: A sign used on goods that have a specific geographical origin and often possess qualities or a reputation that are due to that place of origin.

Infringement of GI: Unauthorized use of a geographical indication by another person

Questions

1. **Define Geographical Indication and describe main features of GI.**
2. **What are the advantages of Geographical Indication?**
3. **Explain briefly disadvantages of geographical Indication.**
4. **Describe geographical indication on the basis of following points.**
   a. Period of protection
   b. Legal Implications
   c. Work to be protected
   d. Infringement
5. **Describe the procedure of registering geographical indication.**
6. **Explain briefly the differences between geographical indication and trademarks.**

7. **Give a brief comparison of Geographical Indication and patent in tabular form.**

8. **Write short notes on the following:**
   a. Merits of geographical indication
   b. Demerits of geographical indication.

9. **What are requirements for registration of geographical indication ?**

10. **Describe role of geographical indication with suitable examples.**

11. **Write Short Notes on:**
    (i) Agricultural GI (ii) Manufactured GI

Chapter 8

Trade Secret

Introduction

A "trade secret" is anything [a formula, process, method, mechanism, tool, pattern or device] which the disclosing party desires to keep secret. Trade secrets usually include such things as the manufacturing details for a product, variations or alternative uses. The main points related to trade secret are briefly presented as follows:

(i) Trade secret is a knowledge that is possessed by an individual or institution and kept secret to prevent others from using the knowledge in a competitive manner.

(ii) There is no specific period for trade secret. It may continue life long or for generations together.

(iii) There is no need of registration for trade secret. Trade secret can never be patented, because it does not publicize the knowledge.

(iv) It does not provide opportunity to others for improvement of innovation.

(v) It is not applicable to books, equipments, plant varieties, designs which are openly used.

Main Features

Main features of trade secret in relation to validity, registration,transfer, infringement and control are briefly presented below:

1. **Period:** It can be protected for indefinite period. In other words, it can be protected so long as the idea/formula is kept secret.
2. **Registration:** There is no need of registration for a trade secret.
3. **Validity:** It is valid with the company concerned.

4. **Transfer:** The owner of the trade secret can reveal the secrecy to other person on payment basis. Trade secret is a property right Hence the owner of the trade secret can sell his trade secret to someone else. Moreover, the owner can license his trade secret to others on royalty basis.
5. **Matters Covered:** A trade secret may consist of following things:

   A formula for a chemical compound

   A process of manufacturing, trating opr preserving materials

   A pattern for a machine or other device or

   A list of customers

   Novelty and inventions are not requisite for a trade secret as they are not for patentability.
6. **Control:** The trade secret is controlled by the owner of the company. Trade secrets protect the application of ideas.
7. **Enforcement:** The trade secret becomes effective when the product is put to continuous commercial use without disclosure.
8. **Infringement:** The infringement of the trade secter is possible through reverse engineering.
9. **Rights Protected:** Protection of trade secret provide the benefit of production and marketing only to the owner of the trade secret. The owner of a trade secret has the right to prevent use or dissemination of the trade secret by any one who learned or derived the trade secret from the owner. The owner has no trade secret rights against any one who independently discovers the trade secret.
10. **Requirements:** Any thing confidential and of commercial use can be protected by trade secret.

How to Protect Trade Secret?

A trade secret can be safeguarded in various ways such (i) by obtaining written agreement from the employees, who know about the trade secret, not to disclose the secret to anyone and (ii) by keeping the documents, containing information about trade secret, in the locking system The employer may recover damages against such a person [employee] revealing the secret. The written agreements are of two types *viz.* A) non-disclosure agreements,and (b) non-compete agreements.

Patents and Trade Secret

There are several differences between patent and trade secret which have been presented in Table 8.1.

Trade Secret of Copyright

There are several differences between trade secret and copyrights which have been presented in Table 8.2.

Table 8.1: Comparison of Patent and Trade Secret

Sl.No.	Particulars	Patents	Trade Secret
1.	Requirement	It is valid in all countries	Any thing confidential of commercial use
2.	Period	15-20 years	Indefinite duration
3.	Registration	Compulsory	Not required
4.	Validity	In original country	With the company concerned
5.	Fee	Fees is charged	No Fee
6.	Legal rights	There are legal Fights	No legal rights
7.	Transfer	Admissible	Possible
8.	Items covered	Products/process/invention//innovation	Formula, pattern, device, business methods, source of supply, customers list etc.
9.	Enforcement	After Registration	With commercial use
10.	Control	By inventor	By owner
11.	Infringement	Possible	Not possible

Table 8.2: Comparison of Trade Secret and Copyright

Sl.No.	Particulars	Trade Secret	Copyrights
1.	Requirement	Any thing confidential and of commercial use	Originality and wide utility
2.	Period	Indefinite duration	70-120 years
3.	Registration	Not required	Not compulsory
4.	Validity	With the company concerned	All countries
5.	Fee	No Fee	No Fee
6.	Legal rights	No legal rights	No legal Rights
7.	Transfer	Possible	Possible
8.	Items covered	Formula, pattern, device, business methods, source of supply, customers list etc.	Literary, music, movie etc.
9.	Enforcement	With commercial use	Immediately after release of work
10.	Control	By owner	By Author/creator
11.	Infringement	Not possible	Very common

Advantages of Trade Secret

1. It is kept in strict confidence.
2. In can be protected for indefinite duration.
3. It provides competitive business advantage.
4. It does not require registration.

Disadvantages of Trade Secret

1. It is lost if secret is revealed.
2. The secrecy may be revealed by the employee.
3. Use of trade secret for long period eliminates the possibility of patenting an invention.
4. It prevents further improvement in a secret process by others.
5. The owner has monopoly in secret information.

Summary

A "trade secret" is anything [a formula, process, method, mechanism, tool, pattern or device] which the disclosing party desires to keep secret. Trade secrets usually include such things as the manufacturing details for a product, variations or alternative uses.

Main features of trade secret in relation to validity, registration, transfer, infringement and control have been discussed. Ways and means of protecting trade secret have been discussed. Comparisons of patent and trade secret and trade secret and copyright have been presented in tabular for ready reference. Advantages and disadvantages of trade secret have been outlined.

Glossary

Copyrights: The exclusive rights to reproduce, sell and distribute a work, prepare derivative works and display the work publicly.

Domain Name: The names and words that companies designate for their registered Internet Web site addresses,

Patent: A document granting an inventor sole rights to an invention. It is an official document which grants sole rights to the inventor for manufacturing and marketing his product/process/invention to derive benefits.

Trademark: A trademark can be a word, name, symbol, device or mark which is used to identify and distinguish the goods or services of one company from goods or services of another.

Trade Name: The name under which a company conducts business, or by which its business, product/service are identified. It may or may not be registered as a trademark

Trade Secret: A formula or process or device used in business that is not published or divulged which gives an advantage over competitors

Questions

1. **Define trade secret and describe its main features?**
2. **What are the advantages of trade secret?**
3. **Discuss in brief disadvantages of trade secret.**

4. **Describe trade secret on the basis of following points.**

   a. Period of protection
   b. Legal Implications
   c. Work to be protected
   d. Infringement

5. **What are ways and means of maintaining the trade secret?**

6. **Explain briefly the differences between patent and trade secret.**

7. **Compare trade secret and copyrights in tabular form.**

8. **Write short notes on the following:**

   a. Trade secret
   b. Duration of trade secret
   c. Merits of trade secret
   d. Infringement of trade secret

9. **What is trade secret infringement ? Explain the same with examples.**

10. **What type of knowledge is covered by trade secret?**

Chapter 9

Related Rights

Introduction

Related rights are rights which are similar to authors' right but which are not connected with the actual author of the work. Both authors' rights and related rights are copyrights in the sense of English or U.S. law. Main points about related rights are given below:

1. Related rights are similar to copyrights.
2. Related rights come under primary property rights.
3. Related rights are also known as neighboring rights.
4. Related Rights are covered by Rome Convention, 1961.

Main Features

The main features of related rights in relation to matters covered, period of protection, registration, authorization, enforcement, infringement etc. are briefly discussed as follows:

1. **Matters Covered:** Related rights or neighbouring rights cover: performers rights, the rights of phonogram producers, publishers' rights in typographical arrangements, the rights of broadcasting organizations in relation to their broadcasts, film producers, database creators, photographers and designers. These rights are internationally protected by the Rome Convention for the Protection of Performers, Producers of Phonograms and Broadcasting Organisations signed in 1961. Within the European Union, the rights of film producers (as opposed to directors) and database creators are also protected by related rights, and the term is sometimes extended to include the *sui-generis* rights in semiconductor topologies and other industrial design rights.

2. **Period of Protection:** The period of protection varies with the matters protected by related rights and also with the countries. The period of protection as per Rome Convention for different matters is as follows:

Table 9.1: Period of Protection for different Matters

Sl.No.	Items/Matters	Period (Years)	Date of Commencement
1.	Performers	15	From end of the year in which performance was made
2.	Phonogram producers	20	From end of the year in which phonogram was first published. It is 50 years in European Union.
3.	Broadcasting organizations	20	From the end of the year in which the broadcast was first made
4.	Film Producers	50	From the end of the year in which the film was first published or otherwise made available to the public, or for fifty years from the end of the year in which the master copy was made if the film is not released
5.	Database creators	15	From the "completion" of the database.
6.	Photographers	25	From date of creation
7.	Designers	25	After creation
8.	Semiconductor designers	10-15	Ten years from the first commercial exploitation or for fifteen years from the first creation.

3. **Relation to authors' rights:** Related rights are independent of any authors' rights which might also exist in the work. Hence a CD recording of a song will be concurrently protected by four different copyright-type rights:
   (i) the authors' rights of the composer of the music;
   (ii) the authors' rights of the lyricist;
   (iii) the performers' rights of the singer and the musicians;
   (iv) the producers' rights of the person or corporation which made the recording.
4. **Rights covered:** The owner of have the rights to prevent other to use their creations without their permission. The rights of different owners are briefly presented below.
   (1) **Performers:** Under the Rome Convention, performers have the right to prevent:

   the broadcast or communication to the public of their performance unless this is made from a legally published recording of the performance;

   the fixation (recording) of their performance;

   the reproduction of a recording of their performance.

   (2) **Phonogram Producers:** The producers of phonograms, that is the person who makes the recording rather than the person who performs

has the right to prevent the direct or indirect reproduction of the recording.

(3) **Broadcasting Organizations:** Rome Convention specifies that broadcasting organisations shall have the right to prohibit (or licence):

(i) the rebroadcasting of their broadcasts;

(ii) the fixation (recording) of their broadcasts;

(iii) the reproduction of fixations of their broadcasts;

(iv) the communication of their broadcasts to the public in places where an entrance fee is charged.

(4) **Film Producers:** Film producers have the right to prevent:

the direct or indirect reproduction of the film (either the master copy or copies thereof);

the distribution of the film (or copies thereof) to the public for sale ;

the "making available to the public" of the film.

(5) **Database Creators:** Database creators have the right to prevent extraction and/or re-utilization of the whole or of a substantial part, evaluated qualitatively and/or quantitatively, of the contents of that database.

Similarly photographers and designers have rights to prevent others to use their creations without permission.

5. **Authorization:** The owners of above mentioned property have the rights to authorise or to provide license to others to make use of their related rights on royalty basis. The performer has the right to licence:

(i) the distribution of recordings of their performance, for sale or other transfer of ownership;

(ii) the rental of recordings of their performances, unless there is a compulsory licence scheme in operation;

(iii) the "making available to the public" of their performances, in effect their publication on the internet.

The owner of the phonogram has the rights to licence:

the distribution of their phonograms, for sale or other transfer of ownership;

the rental of their phonograms, unless there is a compulsory licence scheme in operation;

the "making available to the public" of their phonograms, in effect their publication on the internet.

Once a phonogram has been published, the producer cannot prevent its broadcast: an equitable fee for the licence may be either agreed between phonogram producers and broadcasters or imposed by law.

6. **Registration:** Registration is not compulsory. However, registration is provides legal rights to prevent others to make unauthorised use of the

inventions or creations of the owners. The registration is done in the Copyrights Office and prescribed fee is charged.

7. **Enforcement:** The related rights come into force after registration Without registration, there is no protection of related rights.
8. **Infringement:** The unauthorized use of the protected matters is called infringement. The right holders can take legal action against thoe who are using their creations without permission. The penalties can be imposed against damages.

International Protection of Related Rights

Apart from the Rome convention, there are some other treaties which address the question of the protection of related rights. These treaties are as follows.:

1. Convention for the Protection of Producers of Phonograms Against Unauthorized Duplication of Their Phonograms (Geneva Phonograms Convention, 1971).
2. Convention Relating to the Distribution of Programme–Carrying Signals Transmitted by Satellite (Brussels Convention, 1974).
3. Treaty on Intellectual Property in Respect of Integrated Circuits (IPIC Treaty, 1989).
4. Agreement on Trade-Related Aspects of Intellectual Property Rights (TRIPS, 1994).
5. WIPO Performances and Phonograms Treaty (WPPT, 1996).
6. These treaties do not grant related rights, but endorse that each country should provide related rights to the creators of original work. It will help in promoting artistic and creative work.

Copyrights and Related Rights

Related rights are similar to copyrights. However, the are some differences between them. The comparison of copyrights and related rights is presented in Table 9.2.

Advantages of Related Rights

1. Registration is not compulsory.
2. It protects original work.
3. There is no fee for copyright protection.
4. It promotes creative work.
5. It provides benefits to the creators.

Disadvantages of Related Rights

1. Protection is for shorter period than copyrights.
2. It does provide legal rights without registration.

3. Rights are valid only in the country where registered.
4. It may lead to monopoly.

Table 9.2: Comparison of Copyrights and Related Rights

Sl.No.	*Particulars*	*Copyrights*	*Related Rights*
1.	Requirement	Originality and wide utility	Originality
2.	Period	70-120 years	10-50 years
3.	Registration	Not compulsory	Not compulsory
4.	Validity	All countries	In original country
5.	Fee	Fee is charged if registered	Fees is charged
6.	Legal rights	Legal Rights only after registration	Legal Rights only after registration
7.	Transfer	Admissible	Admissible
8.	Items covered	Literary, music, movie etc	Performers, phonogram, database creators, film producers, designers etc.
9.	Enforcement	Immediately after release of work	After production or release or performance
10.	Control	By Author/creator	By inventor
11.	Infringement	Very common	Possible
12.	Covered by	Berne Convention	Geneva Convention

Summary

Related rights is a term in copyright law, used in opposition to the term "authors' rights". The term neighbouring rights is exactly equivalent, and a more literal translation of the original French *droits voisins* Related rights in civil law are rights which are similar to authors' right but which are not connected with the actual author of the work. Both authors' rights and related rights are copyrights in the sense of English or U.S. law.

The main features of related rights in relation to matters covered, period of protection, registration, authorization, enforcement, infringement etc. have been discussed. Comparison of copyrights and related rights has been presented in tabular form. Advantages and disadvantages of related rights have been outlined.

Glossary

Copyright: The exclusive rights granted to author/creator to reproduce, sell and distribute a work, prepare derivative works and display the work publicly.

Neighbouring Rights: Same as related rights

Related Rights: Related rights is a term in copyright law, used in opposition to the term "authors' rights

Related Right Infringement: An unauthorized use of material protected by related rights.

Questions

1. **What do you mean by related rights? What is their significance?**
2. **Describe briefly the advantages and disadvantages of related rights.**
3. **Describe related rights on the basis of following points.**
   a. Matters protected
   b. Registration
   c. Rights covered
   d. Authorization
4. **Describe briefly the matters that are covered by related rights.**
5. **What are the differences between copyrights and related rights? Describe in tabular form.**
6. **Explain briefly the main features of related rights.**
7. **Write short notes on the following:**
   a. Related rights
   b. Infringement
   c. Copyrights
   d. Neighbouring rights
8. **List the matters covered by related rights. Describe their period of protection.**
9. **Explain briefly the following:**
   a. Matters protected under related rights
   b. Relation of related rights with copyrights
10. **What legal actions can be taken against infringement of related rights ?**

Chapter 10

Infringement or Piracy

Introduction

The infringement of Intellectual Property Rights refers to unauthorized copying or use of a protected form of intellectual property. It is an unauthorized use of material protected by copyright, patent or trademark law is called infringement. It amounts to exploitation of another individual or organization's intellectual property rights without lawful authority.

Definition of Piracy

Piracy refers to an unauthorized duplication of goods protected by intellectual property law (*e.g.* copying software unlawfully). It refers to copying or distributing protected works or products without the permission of the true owner/author of an intellectual property.

It is very clear that definitions of both infringement and piracy are similar and convey the same meaning. Hence, these are synonymous and interchangeable terms. However, infringements is in more usage in case of copyright, patents, trademarks and industrial design and piracy is commonly used in case of unauthorized use of software.

Types of Infringement or Piracy

Infringement may be of different types such (i) copyright infringement, (ii) patent infringement, (iii) trademark infringe, (iv) industrial design, infringement, (v) biopiracy, and (vi) software piracy. These are defined as follows:

1. Copyright Infringement

The illegal use of a copyrighted work or violation of the exclusive rights granted by the copyright law is referred to as infringement of copyrights. In other words, when another party besides the copyright owner reproduces a copyrighted work, in whole or in part, without the copyright owner's permission it is called copyright infringement. Infringement occurs when someone copies, distributes, publicly performs or displays copyrighted work without the permission of the artist or author of the work.

2. Patent Infringement

Making, using or selling an invention protected by a patent without the inventor's permission is called patent infringement. In other words, infringement of a utility patent involves the making, using, selling, offering to sell, or importing of a patented product or process without permission.

3. Trademark Infringement

Violation of trade-mark rights through unauthorized use of a trade-mark is called trademark infringement. In other words, it refers to ihe improper use of another's trademark or a confusingly similar mark that creates the appearance of an affiliation with or actually being the other product or service.

4. Industrial Design Infringement

It refers to violation of industrial design rights through unauthorized use of a design or unauthorized use of industrial design without the permission of owner is called industrial design infringement.

5. Bio-piracy

The illegal collection of indigenous plants by corporations who patent them for their own use is called bio-piracy. It refers to:

(i) Unauthorized use of biological resources such as plants, animals, micro-organisms and genes.;

(ii) Unauthorized use of traditional communities' knowledge on biological resources,

(iii) Unequal share of benefits between a patent holder and the indigenous community whose resources and knowledge have been used; and

(iv) Patenting of biological resources with no respect to patentable criteria such as novelty, non-obviousness and usefulness.

Bio-piracy commonly operates through the application of Intellectual Property Rights to genetic resources and traditional knowledge. Bio-prospecting is the search for biological resources and related indigenous knowledge primarily for the purpose of commercial exploitation. The issue of bio-piracy is mostly raised by under-developed biodiversity rich countries such as India, Brazil and Malaysia among others.

6. Software Piracy

Software piracy is the unauthorized duplication of computer software. In 2003 the estimated losses due to software piracy exceeded 29 billion dollars. Software piracy is of many forms such as (i) end user piracy, (ii) client-server overuse

(iii) internet piracy, (iv) hard disk loading, and (v) software counterfeiting. These forms are briefly discussed as follows.

(i) End User Piracy

This occurs when a company employee reproduces copies of software without authorization. End user piracy can take the following forms:

(i) Sing one licensed copy to install a program on multiple computers;

(ii) Copying disks for installation and distribution;

(iii) Taking advantage of upgrade offers without having a legal copy of the version to be upgraded;

(iv) Acquiring academic or other restricted or non-retail software without a license for commercial use;

(v) Swapping disks in or outside the workplace.

(ii) Client-Server Overuse

This type of piracy occurs when too many employees on a network are using a central copy of a program at the same time. If you have a local-area network and install programs on the server for several people to use, you have to be sure your license entitles you to do so. If you have more users than allowed by the license, that's "overuse".

(iii) Internet Piracy

This occurs when software is downloaded from the Internet The same purchasing rules should apply to online software purchase as for those bought in traditional ways. Internet piracy can take the following forms:

1. Pirate websites that make software available for free download or in exchange for ploaded programs;
2. Internet auction sites that offer counterfeit, out-of-channel, infringing copyright software;
3. Peer-to-Peer networks that enable unauthorized transfer of copyrighted programs.

(iv) Hard-Disk Loading

This occurs when a business who sells new computers loads illegal copies of software onto the hard disks to make the purchase of the machines more attractive. The same concerns and issues apply to Value Added Resellers (VAR) that sell or install new software onto computers in the workplace.

(v) Software Counterfeiting

This occurs when pirates deliberately and illegally duplicate and sell copyrighted material, often making it appear to their customers that they are purchasing an authentic product. When it comes to packaged software, like that sold by Adobe, it is common for pirates to market and sell counterfeit copies of the software CDs along with reproduced packaging, manuals, license agreements, labels, registration cards, and security features.

Illegally copied and sold software is not eligible for support, training, or upgrades. You may not be able to register it, so it may not work properly. And if a vendor is willing to sell you something they don't have the right to sell, you have to wonder what else might get installed on your computer with the software you think you're getting

Penalty Against Infringement or Piracy

The penalties can be imposed for infringement of copyrights, patents, trademarks, industrial design and piracy of software.

1. **Copyrights:** Penalties for copyright infringement can include:

   An injunction against further infringement such as an order preventing the infriger from further copying or distribution of the copyrighted work.

   Impounding or destruction of infringing copies.

   Charging Damages: Either actual damages and the infringer's profit or statutory damages are charged.

   Costs and attorney's fees

   A copyright owner can only sue for infringement on a work that was registered with copyright office. The action against has to be taken within three months of first publication.

2. **Patents:** A patent owner may recover money damages in the form of a reasonable royalty. If the infringement is found willful, three times money is recovered.
3. **Trademarks**: Same action as in case of patents can be taken against infringement of trademarks.
4. **Industrial Design**: The owner has the right to take legal action against the infringer and claim damages. However, the legal action must be taken within three years of the alleged offence.
5. **Software:** There are significant financial and legal penalties for software piracy. Illegal distribution and use of software can result in prison terms of up to five years, and heavy fines [up to US$250,000 in USA]. In civil litigation against those who illegally distribute Adobe software, Adobe can obtain the higher of its lost profits, the infringer's profits, or statutory damages plus attorney fees. But the costs are not measured in money alone When organizations are legally prosecuted for software piracy, they also suffer significant damage to their reputation.

Summary

An unauthorized use of material protected by copyright, patent or trademark law is called infringement Piracy refers to an unauthorized duplication of goods protected by intellectual property law (*e.g.* copying software unlawfully).

It is very clear that definitions of both infringement and piracy are similar and convey the same meaning. Hence, these are synonymous and interchangeable terms. However, infringements is in more usage in case of copyright, patents, trademarks and industrial design and piracy is commonly used in case of unauthorized use of software.

Infringement may be of different types such (i) copyright infringement, (ii) patent infringement, (iii) trademark infringe,(iv) industrial design, infringement, (v) bio-piracy, and (vi) software piracy. All these have been amply discussed.

Software piracy is of many forms such as (i) end user piracy, (ii) client-server overuse, (iii) internet piracy, (iv) hard disk loading, and (v) software counterfeiting.

The penalties that can be imposed for infringement of copyrights, patents, trademarks, industrial design and piracy of software have been discussed.

Glossary

Bio-piracy: The illegal collection of indigenous plants by corporations who patent them for their own use.

Bio-prospecting: Bioprospecting is the search for biological resourcesand related indigenous knowledge primarily for the purpose of commercial exploitation.

Client-Server Overuse Piracy: This type of piracy occurs when too many employees on a network are using a central copy of a program at the same time.

Copyright Infringement: The illegal use of a copyrighted work

End User Piracy: This occurs when a company employee reproduces copies of software without authorization.

Hard Disk Loading: This occurs when a business who sells new computers loads illegal copies of software onto the hard disks to make the purchase of the machines more attractive.

Industrial Design Infringement: Unauthorized use of industrial design without the permission of owner.

Infringement: An unauthorized use of material protected by copyright, patent or trademark law.

Internet Piracy: This occurs when software is downloaded from the Internet.

Patent Infringement: Making, using or selling an invention protected by a patent without the inventor's permission.

Software Counterfeiting: This occurs when pirates deliberately and illegally duplicate and sell copyrighted material.

Software Piracy: Unauthorized duplication of computer software.

Trademark Infringement: Violation of trademark rights through unauthorized use of a trade-mark.

Questions

1. **Define Infringement and describe briefly infringement of copyrights.**
2. **What is piracy? Explain briefly piracy of software.**
3. **Define piracy and discuss briefly different types of software piracy**
4. **What do the following terms signify?**
   a. Bio-piracy
   b. Bio-prospecting
   c. Patent Infringement
   d. Trademarks Infringement
5. **Describe briefly different types of Infringements.**
6. **Write short notes on the following:**
   a. End-use piracy
   b. Internet Piracy
   c. Hard Disk Loading
   d. Software Counterfeiting
7. **Differentiate between the following:**

a. Software piracy and bio-piracy

b. Copyright and patent Infringements

8. **Describe briefly penalties imposed for infringement of copyrights, patents and trademarks.**
9. **List different types of Infringements. Describe any two of them in detail.**
10. **Describe penalties imposed for piracy of software and copyrights.**

Chapter 11

Domain Name

Introduction

Domain refers to the unique name that identifies an Internet site. Domain names have two or more parts separated by dots. For example www.ask-edi.com or www.bcbsks.com The term **domain name** has multiple related meanings:

1. It identifies a web site on the internet (*e.g.* company.com); also referred to as a URL.[Uniform Resource Locator].
2. It allows reference to Internet sites without knowing the true numerical address.
3. It refers to the official name of a computer connected to the Internet. (*e.g.* "www.microsoft.com").
4. It is the unique name of a computer on the Internet that distinguishes it from other systems around the world.
5. They are sometimes colloquially (and incorrectly) referred to by marketers as "web addresses".

Guidelines for Choosing a Domain Name

There are guiding lines which should be given due importance while selecting a domain name. Some tips for choosing a domain name are as follow:

1. **Smaller size:** The domain should not be too long. The domain name can be registered up to 63 characters long, but it is not necessary to use all 63 characters. In the case of domain names the smaller the better. Of course short domain names are harder to find but if you can find one it has a number of advantages such as, it's easier to brand, easier to remember and it's harder to misspell.
2. **Easy to remember:** Try to get a name that's easy to remember and sounds good. General words are best for this but since these are all gone at this

stage you should look for non general names or combinations of general names which rhyme and sound good together. Take elephantdroppings.com for example that's an easy to remember name.

3. **Relate to your business:** If you use keywords which are related to your business industry you will help your search engine rankings. What's even better is if you can get your actual business name or a name similar to your company name as your domain name, this way all your off-line branding will pay off online as Internet users should be able to guess your domain name from your business name.
4. **Avoid similarities:** This means you should avoid choosing domain names that already have similar counterparts online, if you choose one of these people could end up on your competitors site rather than yours because they got confused over the two domain names.
5. **com is best:** You should always aim for the.com version of a name, that's what most people will go to after doing a search or after typing something directly into their address bar, if you register.tv or.net or any of the other available extensions and your competition has the.com version then you've just lost a potential customer.
6. **Use of hyphen:** Use hyphens if your name is more than 3 words long (unless your name includes 3 short words). A name like carbuying.com doesn't really need a hyphen because it's readable the way it is However a name like used-car-buying-scams.com looks better than usedcarbuyingscams.com since there are so many words in the name.
7. **Use of multiple extensions:** Protect your name by registering multiple extensions (.net,.org, etc.) This deters people from copying your name For example, if you own garden-tips-for-us.com you can also register garden-tips-for-us.net and any other extension that's available.
8. **Other points:** The following points should also be kept in mind in choosing a domain name.
   (i) Choose an appropriate top-level domain or domains.
   (ii) Try for a variant of the name of your service or Web site.
   (iii) Avoid names that are too narrow in scope (*e.g.*, the name of a short term project or of one event).
   (iv) Use Web-based wizards or other tools, such as DomainConquest, to look for unclaimed domain names that match key words.
   (v) Avoid trademarks (even your own name, if it is someone else' trademark).
   (vi) Avoid made-up words and alternate spellings (*e.g.*, "pythagoris") (they can be annoying and hard to remember), unless they are well established (*e.g.*, the name of your company).
   (vii) Use at least one reasonably rare word (you have a better chance that the name is not already taken).

(viii) Avoid top-domain-like names (this can create problems with name resolution).

(ix) Don't use an embarrassing name or a name that may antagonize the wrong people.

(x) Don't use numeric digits at the beginning of the name (*e.g.*, "525").

(xi) Don't use non-alphanumeric characters. Most of them don't work over the Internet. Hyphens and underscores are technically OK, but may be difficult for users to remember correctly. A name should never begin with "xn—", which is reserved for encoding Internationalized Domain Names.

(xii) Don't expect case to be preserved.

(xiii) For a large company, consider registering many variant forms and names of products, to avoid their use by competitors, critics, and purveyors of goods with which the company does not wish to be associated.

Main Points

Main points related to domain names such as registration, registration fee, period of registration, enforcenent, authorization, infringement etc are briefly discussed below:

1. Registration

The process through which individuals and organizations obtain a domain name is called registration. Registration of a domain name enables the individual or organization to use that particular domain name for a specified period of time, provided certain conditions are met and payment for services is made. The important points about registration are listed below:

(i) Domain Names can be registered through many different companies known as registrars. The registrar you choose will ask you to provide various contact and technical information that makes up the registration. The registrar will then keep records of the contact information and submit the technical information to a central directory known as the "registry." This registry provides other computers on the Internet the information necessary to send you e-mail or to find your web site. You will also be required to enter a registration contract with the registrar, which sets forth the terms under which your registration is accepted and will be maintained.

(ii) The application is to be filed in a prescribed format called registration forms. These forms are used to register new domain names, new contacts for domain names, and new hosts (name servers) as well as to update domain name, contact, and host records.

(iii) The prescribed fee is to be deposited. It can be deposited by credit card, demand draft or in cash

(iv) The domain name is allotted and activated after the payment of fee is received in the registrar office.

(v) Generic top level domains are available for registration on a global basis.

2. Registration Fee

It refers to the charge for registering, a domain name. The registration fee covers the cost of processing the initial registration and maintaining the domain name record for two years. The registration fee varies from country to country and company to company. After the initial two-year period, the domain name is subject to renewal (re-registration) and the renewal (re-registration) fee will be due on an annual basis if the registrant wishes to renew the domain name's registration.

3. Period of Registration

A domain name is registered for a period of one year which has to be renewed every year. In other words, the domain name is renewed on yearly basis.

4. Enforcement

The domain name becomes operative immediately after registration. It remains active upto 30 days after the expiry of the registration date.

5. Transfer

It refers to the process of changing the party whom is listed as the domain name registrant. The party taking over the domain name is responsible for paying a new registration fee. Domain name transfers are processed using the Registrant Name Change Agreement (RNCA).

6. Infringement

The use of a domain name after the expiray of 30 days from the due date of registration amounts to infringement. It invites legal action against the unauthorized users.

Domain Name System (DNS)

A local database that translates domain names to Internet Protocol (IP) addresses is called domain name system. It is much easier for humans to remember and use names than IP numbers, which are what computers need to find each other on the Internet. Individual Internet computers around the globe maintain their specific portion of this database, and the data held in each portion of the database is made available to all computers and users on the Internet. The DNS comprises computers, data files, software, and people working together.

Parts of a Domain Name

Domain Names always have 2 or more parts separated by the dots. The leftmost piece is the most specific part while the right part is the TLD or Top Level Domain. Generally, there are three parts of a domain name such as (i) top level domain (ii) second level domain, and (iii) third level domain. These are defined as follows:

(i) Top Level Domain (TLD)

The rightmost part of a domain name is called Top Level Domain TLD]. In other words, in a domain name, that portion of the domain name that appears in the rightmost position is known as top level domain. For example, the com in DomainMart.com. The main points related to TLD are given below:

(i) It is the rightmost part of domain name.
(ii) Every domain name ends with top level domain [TLD].
(iii) It is also known as first level domain
(iv) It generally consists of three generic characters.

In the beginning there were only 5 Global top level domains [TLD's] that were recognized by everyone. These TLD's are given below:

.com This TLD is intended for Commercial business around the world.
.net This TLD is historically and is still used by Internet Service Providers.
.org This TLD is intended to search the noncommercial community.
.mil This TLD is restricted to use by the United States Military.
.gov This TLD is restricted to use by United States Government Agencies.

(ii) Second Level Domain

In the Domain Name System (DNS), the next highest level to the left position of the top level domains is referred to as second level domain. In other words, it is that portion of the domain name that appears immediately to the left of the top level domain. For example, the DomainMart in DomainMart.com.

(iii) Third Level Domain

In the Domain Name System (DNS, the next highest level of the hierarchy underneath the second level domains is known as third level domain. It is that portion of the domain name that appears two segments to the left of the top level domain. Third level domains are not the portion of an email address that appears in front of the @ symbol - for example, the webmaster in support@DomainMart.com is not a third level domain.

There can be **fourth and fifth level domains** and so on, with virtually no limitation. An example of a working domain with five levels is *www.sos.state.oh.us*. Each level is separated by a dot or period symbol between them.

Beside these, there are two more terms *viz.* generic top level domain, country code top level domain, subdomain and host nane. These are defined as follows.

Generic Top-Level Domain (gTLD)

It is a top level domain of an Internet address used by a particular class of organization. These are three or more letters long, and are named for the type of organization that they represent The generic top level domains include.com,.net,.org,.mil,.gov, etc. In addition, seven new gTLDs were also selected by ICANN (the

Internet Corporation for Assigned Names and Numbers) on November 16, 2000. These are:.aero (for the entire aviation community);.biz (for business purposes);.coop (for cooperatives);.info (unrestricted);.museum (for museums);.name (for personal names);.pro (for professionals).

Country Code Top Level Domain [ccTLD]

A ccTLD is a country code top-level domain, for example:.mx for Mexico. These ccTLDs are administered independently by nationally designated registration authorities. There are currently 243 ccTLDs reflected in the database of the Internet Assigned Numbers Authority (IANA). WIPO, which has a ccTLD Program, has launched a database portal, facilitating online searches for information related to country code top level domains.

(a) Subdomain

Generally known as a "Domain within a Domain", subdomains are web addresses built upon an existing domain name. In other words, a subdomain is a subdivision of a larger domain. For example, "mail.yahoo.com" is a subdomain of "yahoo.com."

(b) Hostnames

The hostname of an Internet Address is all of the left most pieces of a full internet address if there are more than 2 parts to the address. If there are only 2 parts of the address, then the hostname is equivalent to the domain name. Some examples are as follows.

Full Address	*Hostname*	*Domain Name*	*TLD*
www.keycomputer.com	www	keycomputer.com	com
keycomputer.com	keycomputer.com	keycomputer.com	com
www.google.net	www	google.net	net
my.www.keycomputer.com	my.www	keycomputer.com	com

(i) It is clear from the examples the hostname can be many different things.

(ii) If the entire address consists of only 2 parts, then the hostname is equivalent to the domain name.

(iii) If the entire address consists of more than 2 parts, then the hostname is equivalent to all the left most pieces before the right 2 parts (or the domain name).

In other words, a hostname refers to a domain name that has one or more associated Internet Protocol [IP] addresses. For example, the en.wikipedia.org and wikipedia.org domains are both hostnames, but the org domain is not.

Types of Domain Name Extension

A domain name extension is also called a top level domain name such as.com, org,.net, and a lot more. It is the top most part of the web site address. The domain name extensions are mainly of three types, *viz.* (i) International domain name extension, (ii) National domain name extension, and (iii) alternative domain name extension. These are explained as follows.

1. International Domain Extensions

International Domain Extensions are also referred to as common domain extensions. Three domain name extensions, *viz.*, .com,.net, and.org are the most common on the international front. Moreover, these domain names are called as the "Big 3" because they accessible to anyone, without limitations. They are now widely recognized in all parts of the globe.

2. National Domain Extensions

National Domain Extensions are also known as Country Code Domain Extensions. Now most countries now have a matching top level domain name, under which domain names are sold. The most common domain name extensions are based on two characters ISO country codes. Most popular country top level domains include the.fr for France,.jp for Japan, and.co.uk. These domain name extensions are generally under the control of the government and have high level of stability and durability. The Country Code Top Level Domains [ccTLD] are becoming more popular due to the reducing supply of the more common.com,.net and.org extensions.

3. Alternative Domain Extensions

In the year 2000 the ICANN [Internet Corporation for Assigned Names and Numbers]approved a new set of seven top domains such as.aero,.biz,.coop,.info,. museum,.name, and.pro. The uses of some generic top level domains are explained as follows:

Table 11.1: Uses of some Generic Top Level Domains

Sl.No.	Domain Extension	Meant for or Designated for
1.	.com	For the commercial organizations.
2.	.net	For network services like Internet Service Providers.
3.	.org	For non-profit organizations.
4.	.aero	For entire aviation community
5.	.biz	For business purposes
6.	.coop	For cooperatives
7.	.inf	For unrestricted use
8.	.museum	For museums
9.	.name	For personal names
10.	.pro	For professionals

They are generally called "alternative top level domain names". They are officially known, thus will work in all parts of the globe, in any browser or other software. These are also known as generic top level domains.

Recently, ten new domain name extensions have been included. These are are. asia,.cat,.jobs,.mail,.mobi,.post,.tel,.travel, and.xxx.

Relative Significance

There are many types of domain extensions and each has its importance and value The importance of some domain extensions is given below:

(i) **".com"** for example is perceived in the public as more credible and usable than a ".tv" or ".mobi", it is more "official." In the academic world, the ".com", is preferred rather ".org" or ".edu"**.com** These sites are generally businesses of some sort or informational portals. The most popular extension, this is the one people know about.

(ii) **.org, edu** These sites are more credible and usually often are dedicated to a school or non-profit organization.

(iii) **.tv, .mobi, .us,** etc These extensions generally are different types of domain extensions associated with how you access them.**mobi**, for example, refers to accessing it through your mobile phone. They are not as hot as a.**com**, though still very popular and growing.

Role of ICANN

ICANN refers to the Internet Corporation for Assigned Names and Numbers ICANN is responsible for managing and coordinating the Domain Name System (DNS) to ensure that every address is unique and that all users of the Internet can find all valid addresses. It does this by overseeing the distribution of unique IP addresses and domain names. It also ensures that each domain name maps to the correct IP address.

ICANN is also responsible for accrediting the domain name registrars "Accredit" means to identify and set minimum standards for the performance of registration functions, to recognize persons or entities meeting those standards, and to enter into an accreditation agreement that sets forth the rules and procedures applicable to the provision of Registrar Services.

ICANN's role is very limited, and it is not responsible for many issues associated with the Internet, such as financial transactions, Internet content control, spam (unsolicited commercial email), Internet gambling, or data protection and privacy

Trademarks and Domain Names

There are some similarities and some differences between trademarks and domain names which presented in Table 11.2.

Advantages of Domain Names

There are several advantages of domain names. Some advantages of domain names are as follows.

1. **Naming Rights:** A domain name may only be registered once and for one buyer only. Since it is in a first come, first served basis, the first one to

Table 11.2: Comparison of Trademark and Domain Name

Sl.No.	*Particulars*	*Trademarks*	*Domain Names*
A.	**Similarities**		
1.	Registration	Essential	Essential
2.	Types	Many	Many
3.	Control	By owner	By owner
4.	Registration fee	Payable	Payable
5.	Renewal	Possible	Possible
6.	Transfer	Possible	Possible
7.	Requirement	Distinct	Distinct
8.	Infringement	Possible	Possible
B.	**Differences**		
1.	Registration with	Trademarks office	Registrar Internet
2.	Validity	Within country	Both global and national
3.	Items covered	Identification marks	Internet web site
4.	Enforcement	With commercial use	After registration
5.	Annual Fee	Charged	Charged once

register the name shall have the right to use the name. It is a memorable website address. The owner has complete control over his website.

2. **Identity:** A domain name represents an identity which visitors always recognize. It is just like a brand name or a logo,
3. **Simplicity:** A domain name shortens the URL address of your blog and it is much easier to remember. It is easier for users to remember domain name than host-domain-path which is longer (such as *instruct.uwo.ca/~craven*).
4. **Adds Credibility:** A domain name projects a professional image of a owner and thus adds credibility to the owner. In other words, a domain name looks more professional and does have more perceived credibility from all your visitors.
5. **More Advertisers:** Some advertisers prefer a domain name than a sub domain name. With the use of a domain name, there is an expansion of opportunities for your blog. More advertisers, more tasks.
6. **More impressive:** It is very impressive (looks less like a personal site)
7. **Transportable:** It is transportable to another hosting service.
8. Customer service is readily available.
9. Domain name makes the keywords clear to search engines because the words are separated.

Disadvantages of Domain Names

1. It must be registered with appropriate service, normally for a small yearly fee.

2. Sometimes the name we want may already be taken; we may have to pay a lot (or sue) to obtain it.
3. One may lose the domain name unexpectedly because of registrar incompetence.
4. It requires proper maintenance of DNS records on a server.

Summary

Domain refers to the unique name that identifies an Internet site. Domain names have two or more parts separated by dots. For example www.ask-edi.com or www.bcbsks.com The term **domain name** has multiple related meanings which have been explained:

Main points related to domain names such as registration, registration fee, period of registration, enforcenent, transfer, infringement etc have been discussed.

There are three parts of a domain name, *viz.* (i) top level domain, (ii) second level domain, and (iii) third level domain. These have been defined. Other terms such generic top level domain [gTLD], country code top level domain [ccTLD] subdomain and hostname have also been defined.

A domain name extension is also called a top level domain name such as.com,. org,.net, and a lot more. It is the top most part of the web site address. The domain name extensions are mainly of three types, *viz.* (i) International domain name extension, (ii) National domain name extension, and (iii) alternative domain name extension. These have been explained. Country code top level domains for different countries are presented in tabular form.

The role and limitations of ICANN have been discussed. A comparison of trademarks and domain names is given in tabular form. Advantages and disadvantages of related rights have been outlined.

Glossary

Domain Names: A domain name is also called a Web address. A domain name is used as an address for your Web site and e-mail, just like www.janesbagels. com and orders@janesbagels.com.

Domain Pointer/Sub-domains: Your hosting account will allow you to point or assign several domains (*e.g.* janesbagels.com) to your hosting package, including domains that you have registered with Network Solutions and domains that you have registered with other providers. You can also create sub-domains for these registered domains.

Domain Protect: Domain Protect adds an extra layer of protection to your domain name registration, guarding it against unauthorized domain name transfers.

Extensions: Network Solutions offers a variety of domain name extensions. Protecting brand identity has become very important, so often customers will register multiple extensions and variations of their domain names.

Extranet: An extranet is a network that allows a company to share information with other businesses and customers.

Register Domain: To establish an Internet identity you need to have a domain name. In order to do this, you'll need to register a domain name. To do so, you can go to www.networksolutions.com and type your desired name in the search box and purchase an available domain.

Domain Name System (DNS): A local database that translates domain names to Internet Protocol (IP) addresses.

A generic top-level domain (gTLD): It is a top level domain used by a particular class of organization. These are three or more letters long, and are named for the type of organization that they represent (for example,.com for commercial organizations).

Internet Protocol (IP) Numbers (IP addresses): A unique, numeric identifier used to specify hosts and networks.

Internet Service Provider (ISP): A person, organization, or company that provides access to the Internet.

Premium Domain Name: Top level domain names, or also known as premium domain names, are commonly knows as a domain ending with the extension ".com". The.com is the most recognized, due to its time longest on the Internet and its commercial value. Moreover, .com domain names are of much higher quality and offer a more competitive advantage with the search engines and competition.

Registration: The process through which individuals and organizations obtain a domain name.

Registration Fee: The charge for registering, a domain name. InterNIC' registration fee covers the cost of processing the initial registration and maintaining the domain name record for two years.

Registry: A registry is responsible for delegating Internet addresses such as Internet Protocol (IP) numbers and domain names, and keeping a record of those addresses and the information associated with their delegation.

Renewal (Re-registration): The process of reinitiating a domain name's registration for a specified period of time.

Renewal (Re-registration) Fee: The charge for renewing an existing domain name registration. InterNIC' renewal (re-registration) fee covers one year of maintenance for the domain name record and is assessed each year on the anniversary of the original registration.

Renewal (Re-registration) Notice: A notice sent to registrants 60 days before the anniversary of their registration date to let them know that their domain name will be due for renewal (re-registration) and that an invoice will be sent for the renewal (re-registration) fees.

Second Level Domain: In the Domain Name System (DNS), the next highest level of the hierarchy underneath the top level domains. In a domain name, that portion of the domain name that appears immediately to the left of the top level domain. For example, the DomainMart in DomainMart.com.

Sub-domain: Generally known as a "Domain within a Domain", sub-domains are web addresses built upon an existing domain name. In other words, a sub-domain is a subdivision of a larger domain. For example, "mail.yahoo.com" is a sub-domain of "yahoo.com."

Third Level Domain: In the Domain Name System (DNS, the next highest level of the hierarchy underneath the second level domains. It is that portion of the domain name that appears two segments to the left of the top level domain.

Top Level Domain (TLD): In the Domain Name System (DNS), the highest level of the hierarchy after the root. In a domain name, that portion of the domain name that appears furthest to the right. For example, the com in DomainMart.com.

Transfer: The process of changing the party whom is listed as the domain name registrant.

URL (Uniform Resource Locator): A pointer to a page on the WWW or other resource on the Internet. A URL starts with the "http://" for web-pages, or "ftp://" for FTP. That's followed by the domain name (www.DomainMart.com), or the path to a document such as this (http://www.DomainMart.com/news/Glossary_Terms.htm).

Questions

1. **What do you mean by domain names? What is their significance?**
2. **Describe briefly the advantages and disadvantages of domain names.**
3. **Describe related domain names on the basis of following points.**
   a. Registration
   b. Registration Fee
   c. Renewal of registration
   d. Authorization
4. **Describe briefly the types of Internet domains.**
5. **Explain briefly the differences between domain name and trademark.**
6. **Explain briefly the main features of domain names.**
7. **Write short notes on the following:**
   a. First Level Domain
   c. Second Level Domain
   c. Third Level Domain
   d. Internet Protocol
8. **Give the full form of the following:**
   a. DNS
   b. URL
   c. RNCA
   d. TLD
   e. ICANN

9. **What do the following terms signify. ?**

   a. Top Level Domain
   b. National domain
   c. International domain
   d. Internet server

10. **What points should be taken into account while selecting a domain name?**

11. **Describe the qualities of a good domain name.**

12. **Give a brief account of Domain Name extensions.**

13. **Write short notes on the following:**

    (i) Generic Top Level Domain [gTLD]

    (ii) Country Code Top Level Domain [ccTLD]

14. **What is ICANN? Describe its roles and limitations.**

Chapter 12

Trade Name

Introduction

A trade name is the name under which a product is commercially known. The commercial name by which a chemical is known is called trade name. One chemical may have a variety of trade names depending on the manufacturers or distributors involved Main points related to trade name are given below.

1. It is also known as business name, trading name, assumed name, brand name and corporate name.
2. It may or may not be registered.
3. It distinguishes a particular business from others.
4. It can be exclusive or non-exclusive.
5. It is generally used on letter heads and bank accounts.
6. It is the name given to a particular substance by each company that manufactures it.
7. It is the name under which a company conducts its business.

It is s the business name of the person or organization making and/or selling a particular product. "General Motors, Inc" is the "trade name" of the company.

Choosing a Business Name

There are a number of considerations in choosing a business name, and securing it for use in a trading for a company, partnership or sole trader. A business name is the name by which a sole trader, partnership, limited liability partnership or a company. A business name must have distinctive, descriptive and legal elements and each element is considered as part of the whole name. The name should not be too general. These are briefly explained below:

1. **Distinctive:** This is the element which is the unique identifier of the business. This element is the main identifier of your business name. This is the part of the name that sets it apart from other businesses of the same type. **Every business name must have a distinctive element.** A word by itself may not be distinctive but it would be distinctive when used in combination with other words.
2. **Descriptive:** This is the element which describes the main aspect of the business; and
3. **Legal:** This is the element when the name is used by a corporation. The legal element indicates the status of the organization as a corporate body.
4. **It should not be too General:** Names that are too general usually lack the distinctive element. They infringe on existing registered names, thus leading to confusion among the public. Names which are too general will also limit the number of names available to the public in the future.

Moreover, a business name should not suggest or imply a connection with any government or government department A business name must not suggest a connection with the Royal Family The name of a society must be such that it will not be confused with existing names of societies, corporations, partnerships or trade names. When choosing a name for a society, all of the other naming guidelines should be followed.

Types of Business Ownership

There are three types of business ownership or business structures which are most commonly used. These are Sole Proprietorship, Partnership, and Company or Corporation. A brief account of these business structures with regard to taxation, liability, duration and registration is presented below:

1. Sole Proprietorship

is the simplest form of business in which one person performs all the functions that are required for the successful operation of a business. The proprietor secures the capital, hires any employees, establishes and operates the business, assumes all the risks, accepts all profits or losses and pays all the taxes. The proprietor is said to be self-employed.

Taxation: Any income or losses are claimed on the owner's personal tax return each year. Business deductions are permitted.

Liability: Liability arising out of the business activity is borne entirely by the proprietor. All assets, whether business or personal, can be legally taken and used to discharge the liability.

Registration: A sole proprietorship can be operated under the individual's name without the need for registration. If a business name is used such and company the registration is required.A business name registration is valid for three year from the date of registration. At that time, registration will expire and renewal of registration must be completed by the sole proprietor of the business name or the registration will be removed.

Duration: The business ends with the death or incapacity of owner. Therefore, this type of business cannot be inherited. However, the individual assets can be used to carry on a business. The business name can be removed when a sole proprietor discontinues business.

2. Partnership

A partner-ship is a voluntary contract between two or more competent persons to place their money, efforts, labour and skill in business with the understanding that there shall be a proportional sharing of profits and losses between them. A partnership is governed by the *Partnership Act*. For every partnership, there should be a written partnership agreement between the partners setting out his respective right, interest and responsibility.

Taxation: A Partnership Agreement can allocate the profits or losses in any ratio agreed to between the partners but if there is no Agreement, the profits must be allocated equally. Business deductions are taken by the partnership before the income is distributed to the partners and claimed on their personal tax returns.

Liability: Partners are personally liable for debts of the partnership and the full amount of these debts can be collected from one or more of the partners rather than the debt being equally shared. Partners can also be held liable for acts committed by one of their partners in the normal course of business.

Registration: A declaration of the partnership must be registered under the Partnership Act *w*ithin three months of the formation of a partnership, A computerized name search is made of the proposed partnership name prior to registrationA partnership registration is valid for three years from the date of registration. At that time, the registration will expire and a renewal of registration must be completed by the partners or the partnership registration will be removed.

Duration: This business organization ends with death, incapacity, withdrawal or bankruptcy of any partner, unless otherwise agreed to in a Partnership Agreement. When a partnership discontinues business, the business name can be removed from the business names database.

3. Corporation

A corporation or company is a separate legal entity having an existence separate and apart from the business person that created it. A company can be formed under the Company Act or the Corporations Act. The provincial corporations are most common. Its continuance depends only on the filing of annual returns and annual meetings of shareholders and directors. The Company Act allows a company to be owned by just one person. If more than one person is involved in the business, a Unanimous Shareholder's Agreement is recommended to establish the operational rules between the shareholders including dispute resolution and share evaluation. It is wise to consult with a lawyer to ensure that the company is set up in a manner that meets your business' present and future needs.

Taxation: The Company pays taxes on corporate earnings after the business deductions are made. Money taken out of the company in the form of share dividends or wages by the shareholders is claimed on their personal tax returns.

Liability: Liability is limited to the assets of the company and the shareholders are not personally liable for the debts of the company unless allowed by law. The directors of a company may be held personally liable under certain circumstances, such as unpaid contributions under the Income Tax Act. Most banks and many other large creditors will require the owner of a small company to provide a personal guarantee to secure a loan. These factors reduce the effectiveness of limited liability for small business owners.

Registration: A corporation can register a business name under the *Partnership Act* in the same manner that an individual(s) can register a Sole Proprietorship or Partnership. A corporation may wish to use a name, other than their corporate name, for the conduct of some or all of their business activity.

A business name registration is valid for three years from the date of registration. At that time, registration will expire and renewal of registration must be completed by an officer of the corporation or the business name will be removed. When a corporation discontinues the use of a trade name, the name can be removed from the business names database. The name can be removed on written request only by the registered owners of the business name.

If a business name is used by a corporation, a declaration of this in writing must be filed. An officer authorized by the corporation must complete the declaration on behalf of the corporation. A corporation wishing to register a trade name is required to have a computerized name search of the proposed name prior to registration.

Duration: A company is its own legal person, therefore it survives the death, incapacity or withdrawal of any shareholder or director. Its survival depends on the filing of annual returns and fulfillment of the requirements of the Company Act.

Table 12.1: Comparison of Three Forms of Business Structures, *viz.* Sole Proprietorship, Partnership and Corporation

Sl.No.	*Particulars*	*Sole Proprietors*	*Partners*	*Corporations*
1.	Registration	Not required	Required	Required
2.	Duration of registration	Three years if registered	Three years	Three years
3.	Renewal of registration	After 3 years	After 3 years	After 3 years
4.	Liability	Owner	Partners	Share holders
5.	Ownership	Individual	Partners	Share holders
6.	Transfer	Permitted	Permitted	Permitted
7.	Term terminates	With death of owner	After specified period	Unlimited
8.	Infringement	Possible	Possible	Possible
9.	Rights with	Owner	Partners	Board of Directors
10.	Enforcement	With start of business	After registration	After registration
11.	Validity	In country of registration	In country of registration	In country of registration

Main Features

The main features of trade names in relation to registration, matters covered, duration of protection, rights covered, requirements, authorization, enforcement and infringement are briefly discussed as follows:

1. **Registration:** It may or may not be registered. However, registration is necessary to get legal rights. Trade names are registered by the state in which the person or company is based. For each State, it has to be registered separately. The application is filed in the prescribed form and prescribed processing fee is to be deposited.

   If a person (or persons associated in partnership) is engaged in business for trading, manufacturing, contracting, or mining purposes, he or she must register the business name by filing a declaration with the Registrar through an authorized service provider. The authorized service provider will examine your information and if it meets the current legislative requirements, will process the request and issue you a proof of filing to verify that the registration has occurred.
2. **Matters Covered:** Trade names cover all types of business, *viz.* sole proprietership, partnership; corporate business, etc.
3. **Duration of Protection:** Initially the trade name is registered for a term of three years which is renewable for a further term of three years each.
4. **Rights Covered:** The sole owner of the trade name has all the rights related to his business, partners have equal rights and in a corporation, board of Directors have the rights.
5. **Requirements:** The basic requirements for registering a trade name are that that the name should be new and distinctive. It should not confused with the trade name of any other person or corporation.
6. **Authorization/Transfer:** The owner of the trade name has right to transfer his business to any other person who is interested in the business. For transfer the request is submitted in the prescribed form and transfer fee is to be paid.
7. **Enforcement:** The trade name comes into force immediately after registration and it expires after death of the owner or after its term if not renewed timely.
8. **Infringement:** Unauthorised use of a trade name by another without permission amounts to infringement. The trade name owner has rights to prevent such use. He can take legal action against infringer.

Trademark and Trade Name

There are some similarity and some differences between trademarks and trade names which are presented in Table 12.2.

Table 12.2: Comparison of Trademarks and Trade Names

Sl.No.	*Particulars*	*Trademarks*	*Trade Names*
1.	Governed by	Primary Rights	Primary Rights
2.	Registration	Compulsory	Compulsory
3	Period of protection	Renewable indefinitely	Renewable
4.	Protection over	Names, images and short phrases	Products
5.	Validity	In original country	Country of registration
6.	Enforcement	With commercial use	With start of business
7.	Requirements	New and distinctive	New and Distinct
8.	Transfer	Permitted	Permitted
9.	Infringement	Possible	Possible

A trade name identifies the business itself, whereas trademarks identify goods or services. A trade name can also serve as a trademark if it meets the requirements of a trademark.

Advantages and Disadvantages of Trade Names

Advantages and disadvantages of three types of business names,*viz*. sole proprietorship, partnership and corporations are briefly presented below.

Table 12.3: Advantages and Disadvantages of Sole Proprietorship, Partnership and Corporations

Sl.No.	*Advantages*	*Disadvantages*
A.	**Sole Proprietorship**	Lack of continuity
1.	Low start-up costs	More difficult to raise capital
2.	Greatest freedom from regulation	
3.	Direct control by owner	
4.	Minimum working capital requirements	
5.	Tax advantage to small owner	
6.	All profits to owner	
B.	**Partnership**	
1.	Ease of formation	Unlimited personal liability
2.	Low start-up costs	Lack of continuity
3.	Additional sources of venture capital	Divided authority
4.	Broader management	Difficulty in raising additional capital
5.	Limited outside regulation	Hard to find suitable partners
C.	**Corporation**	
	Limited liability	Closely regulated
	Specialized management	Most expensive to organize
	Ownership is transferable	Charter restrictions

Contd.

Table 12.3–*Contd...*

Sl.No.	*Advantages*	*Disadvantages*
	Continuous existence	Extensive record-keeping necessary
	Legal entity	Double taxation, except when organized as an "S Corporation"
	Easier to raise capital	Difficult to liquidate investment
	Unity of action account having	
	Centralized authority in board of directors	

Business Names vs Corporation Names

There are differences between business names and incorporated names, as well as differences between the resulting entities formed. An incorporated name must contain a legal element, while a trade name or partnership name cannot. Another difference between a business name and an corporation name is that a corporation name is only granted once.

A difference between a registered corporation and a trade name or partnership, is that the trade name or partnership is not a distinct legal entity. Therefore, the proprietor(s) or partners are completely responsible for the debts and liabilities of the business.

Summary

A **trade name**, also known as a **trading name** or a **business name**, is the name which a business trades under for commercial purposes, although its registered, legal name, used for contracts and other formal situations, may be another. Pharmaceuticals also have trade names (*e.g.* "**Aspirin**"), often dissimilar to their chemical names ("**acetylsalicylic acid**"). Trading names are sometimes registered as trademarks or are regarded as brand names.

A business name must have distinctive, descriptive and legal elements and each element is considered as part of the whole name. The name should not be too general.

There are three types of business ownership or business structures which are most commonly used. These are Sole Proprietorship, Partnership, and Company or Corporation. A brief account of these business structures with regard to taxation, liability, duration and registration is presented. A comparison of these business structures is presented in tabular form and their advantages and disadvantages have been outlined.

The main features of trade names in relation to registration, matters covered, duration of protection, rights covered, requirements, authorization, enforcement and infringement have been discussed.

Comparison of trademarks and trade names has been presented in tabular form.

Glossary

Brand Name: Same as trade name.

Business Name: Same as trade name.

Corporate Name: Same as trade name.

Fictitious Name: An assumed name.

Juridical Person: Juridical person refers to a partnership, corporation, cooperative, and legal labor organization, etc. that are registered in the particular concerned agencies.

Natural Person: Natural person refers to sole proprietorship/individual

Trade Name: The name under which a company conducts business, or by which its business, product/service are identified. It may or may not be registered as a trademark.

Trademarks: A symbol used to identify and distinguish companies, partnerships, and businesses, as opposed to marks used to identify and distinguish goods or services.

Trading Name: Same as trade name.

Questions

1. **What do you mean by trade name ? Describe its significance.**
2. **Describe briefly the advantages and disadvantages of trade names.**
3. **Describe briefly trade names on the basis of following points.**
   a. Registration
   b. Period of Registration
   c. Enforcement
   d. Duration
4. **Describe briefly the main features of trade Names.**
5. **Give a comparison of trade names and trademarks in tabular form.**
6. **Write short notes on the following:**
   a. Trade Name
   b. Infringement
   c. Trademarks
   d. Trade secret
7. **Describe briefly the rights covered under trade names.**
8. **Explain briefly the basic requirements for registration of trade name.**
9. **What legal actions can be taken against infringement of trade names ?.**
10. **Describe briefly the following:**
    (i) Rights covered under trade name
    (ii) Matters covered by trade name

11. **Describe sole proprietorship on the basis of liability, taxation, registration and duration.**
12. **Discuss briefly the partnership type of business structure on the basis of liability, taxation, registration and duration.**
13. **Explain briefly the corporation type of business structure on the basis of liability, taxation, registration and duration.**
14. **Give a brief comparison of sole proprietorship and partnership type of business structures.**
15. **Compare sole proprietorship and corporation types of business structures in a tabular form.**
16. **Define sole proprietorship type of business structure and describe its advantages and disadvantages.**
17. **Define partnership type of business structure and describe its advantages and disadvantages.**
18. **Define corporation type of business structure and describe its advantages and disadvantages.**
19. **Write short notes on the following:**
    (i) Sole proprietorship business
    (ii) Partnership type business
    (iii) Corporation type of business
20. **What do the following terms signify?**
    (i) Natural person (ii) Juridical person
    (iii) Brand Name (iv) Assumed Name

Section II

Sui-Generis IP Rights

Chapter 13

Sui-Generis Intellectual Property Rights

Introduction

Sui-generis is a Latin expression, literally meaning of its own kind/genus or unique in its characteristics. In law, it is term of art used to identify a legal classification that exists independently of other categorizations because of its uniqueness or due to the specific creation of an entitlement or obligation. In Intellectual Property there are rights which are known as being sui-generic to owners of a small class of works. Such Intellectual Property Rights include (i) database rights, (ii) mask work, (iii) plant breeders' rights, (iv) farmers' rights, (v) moral rights, (vi) supplementary protection certificate, and (vii) indigenous intellectual property. A brief account of all these is presented in this chapter.

1. Database Rights

Database rights are a form of exclusive right introduced by European Union Law in 1996 in those countries which follow EU Law. A 'database right' is an intellectual property right given to a computer database. The main points about database rights are as follows:

1. In most countries database are covered by copyright law to some degree.
2. The database rights were introduced by European Union Law to provide greater protection to collection of information.
3. Database right is covered by sui-generic rights.
4. **Duration:** Database rights last for 15 years, but can be extended if the database is updated.
5. **Protection:** Database right prevents copying of substantial parts of database. The protection is over the information, not over the form of expression.

Table 13.1 Intellectual Property Law: List of Sui-generic Rights

Sl.No.	*Items of Sui-generic Right*	*Definition or Brief Description*
1.	Database rights	A 'database right' is an intellectual property right given to a computer database
2.	Mask Work	A mask work is a two or three-dimensional layout of an integrated circuit (IC or "chip")
3.	Plant Breeders' Rights	Rights granted to the breeder of a new variety of plant
4.	Farmers' Rights	The legal rights provided to farmers to save, use, sow, replant, exchange, share or sell his farm produce
5.	Moral Rights	Moral rights are a special extension to copyright that gives the originator of a copyrighted work rights over its use.
6.	Supplementary protection certificate	A supplementary protection certificate (SPC) is a sui-generis, patent-like, intellectual property right that is available for medicinal and plant protection products.
7.	Indigenous Intellectual Property	The knowledge, innovations and practices of indigenous people and local communities.

2. Mask Work Rights

A mask work is a two or three-dimensional layout of an integrated circuit (IC or "chip"), *i.e.* the arrangement on a chip of semiconductor devices such as transistors and passive electronic components such as resistors and interconnections. Main points related to mask work are listed as under.

1. Mask work can not be protected under copyright law or patent law.
2. The mask work exclusive rights were first granted in USA by Semiconductor Chip Protection Act of 1984. Such legislations also exist in Canada and Australia.
3. **Duration:** Rights for semiconductor mask work lasts only for two years (if unregistered) or 10 years (if registered).
4. The exclusive rights in mask work are somewhat like those of copyright the right to reproduce the mask work or distribute an IC made using mask work.
5. Reproduction for reverse engineering of a mask work is permitted by law
6. Mask work rights exist when they are created, even without registration
7. Protection under mask work cannot be provided to a mask that is not original.

Database and Copyrights

There are some similarity and some differences between database and copyrights which are presented in Table 13.2.

Table 13.2: Comparison of Database and Copyrights

Sl.No.	Particulars	Database Rights	Copyrights
1.	Governed by	Sui-generic rights	Primary intellectual rights
2.	Registration	Not compulsory	Not compulsory
3.	Period of protection	15 years	70-120 years
4.	Protection over	Information	Expression
5.	Validity	Country of registration	Country of registration
6.	Enforcement	After creation	After creation

Database and Mask Work

There are some similarity and some differences between database and mask work which are presented in Table 13.3.

Table 13.3: Comparison of Database and Mask Work

Sl.No.	Particulars	Database Rights	Mask Work
1.	Governed by	Sui-generic rights	Sui-generic rights
2.	Registration	Not compulsory	Not compulsory
3.	Period of protection	15 years	Unregistered 2 Registered 10 years
4.	Protection over	Information	Integrated Circuits
5.	Validity	Country of registration	Country of registration
6.	Enforcement	After creation	After creation

3. Plant Breeders' Rights

Plant breeders' rights, also known as plant variety rights (PVR), are intellectual property rights granted to the breeder of a new variety of plant. Plant Breeders Rights are granted to novel plant varieties that are distinctive, uniform, and stable (*e.g.* cultivars bred true-to-type for desired traits). The legal protection of a new plant variety is granted to the breeder or his successor. The effect of PBR is that prior authorization is required before the material can be used for commercial purposes.

Main Features

Main features of Plant Breeders' Rights in relation to registration, duration, validity, matters covered, requirements, transfer, control, enforcement, rights protected and infringement. are briefly presented below:

1. **Registration:** Registration is essential to get legal rights. There is no protection without registration. A nominal processing Fee is charged for registration.
2. **Duration:** The period of protection varies with plant species. It is 20 years for field crops, 25 years for trees, vines, fruit trees, ornamental trees and shrubs [UPOV 1991].

3. **Validity:** The protection of the Plant Breeders' Rights. is valid only in the country where it has been registered. The protection in other countries can be obtained by filling separate application in each country.
4. **Matters Covered:** The protection right can be granted for varieties of all botanical genera and species. The variety should have a designation [name] as per the rule of International Code of Nomenclature.
5. **Requirements:** There are four basic requirements for protection of a variety under PBR, *viz.* (i) novelty, (ii) distinctiveness, (iii) uniformity and (iv) stability.
6. **Transfer:** An authorized plant breeder has right to authorize other interested persons for commercial production and marketing of his variety on royalty basis.
7. **Control:** The authorized plant breeder has right to prevent other from commercial production and marketing of his variety without permission.
8. **Enforcement:** The Plant Breeders' Rights come into force immediately after registration of the variety.
9. **Rights Protected:** It provides exclusive rights to the owner for commercial production and marketing of his variety.
10. **Infringement:** Unauthorized production and marketing of a registered variety by other person amounts to infringement. The owner has the right to take legal action against the infringer and claim damages.

Advantages of PBR

1. Breeders get benefit of their variety.
2. PBR help in faster development of seed industry.
3. PBR lead to improvement in quality because of competition.
4. PBR are useful in procurement of good material on payment basis.
5. PBR help in enrichment of genetic resources.

Disadvantages

1. It will promote monopoly.
2. It will encourage unhealthy practices.
3. It may lead to increase in prices.
4. There will be reduction in genetic variability.
5. There will be compulsion to purchase frsh seed every year.

4. Farmers' Rights

Farmers' rights refer to the rights arising from the past, present and future contributions of farmers in conserving, improving and making available plant or animal genetic resources, particularly those in the centres of origin/diversity. In other words, the legal rights provided to farmers to save, use, sow, replant, exchange share or sell his farm produce including seed of a variety protected under Plant Variety Protection Act refer to Farmers Rights. The purpose of these rights is to "ensure full benefits to farmers and support the continuation of their contributions.

The FAO Conference held in Rome from 11-29 November, 1989 endorsed the concept of Farmers Rights with a view to:

(i) Ensuring global recognition of the need for conservation and the availability of sufficient funds for these purposes;

(ii) Assisting farmers and farming communities throughout the world, especially those in areas of original diversity of plant genetic resources, in the protection and conservation of their PGR and of the natural biosphere; and,

(iii) Allowing the full participation of farmers, their communities and countries in the benefits derived, at present and in the future, from the improved use of PGR.

5. Moral Rights

Moral rights are rights of creators of copyrighted works generally recognized in civil law jurisdictions and first recognized in France and Germany, before they were included in the Berne Convention for the Protection of Literary and Artistic Works in 1928. Moral rights are a special extension to copyright that gives the originator of a copyrighted work rights over its use. The originator can specify moral conditions on the use or exploitation of their creation, even when the rights are licensed or sold to others'. The main points related to moral rights are presented below:

1. These are rights provided to authors and directors of copyrights to protect their personal interest in relation to their creations. These rights are provided beyond those strictly recognized in copyright law.
2. Moral Rights were granted to authors under the Berne Convention (1989) giving the author (i) right to attribution, and (ii)) right to integrity.
3. The creator of the work has the right of attribution and integrity, even after sale or transfer of the copyright. In other words, the creator or author shall have the right to claim authorship of the work even after sale or transfer of the copyright.
4. **Transfer:** The copyrights can be transferred or assigned to others, but moral Right cannot be transferred or assigned.
5. The author has right to object to any distortion, mutilation or other modification which would be prejudicial to his honour or reputation.

6. Supplementary Protection Certificate

This term is used in European Countries. This is an extension of the term given to pharmaceutical or plant protection patent. A supplementary protection certificate (SPC) is a sui-generis, patent-like, intellectual property right. This type of right is available for medicinal products, such as drugs, and plant protection products, such as insecticides, and herbicides. Main points related to supplementary protection certificate are given below:

1. It is in use in European Union member countries.
2. It is a sui-generis, patent like, intellectual property right.
3. The SPC comes into force only after the expiry of corresponding patent.
4. **Duration:** It has maximum life time of 5 years when granted after 10 years after filing date of a patent. The term would be less than 5 years if it is granted after 5 years from the date of filing the corresponding patent. However, the market exclusivity cannot exceed 15 years.
5. The SPC provides long term authorization to a patent for marketing.
6. **Conditions of Issue:** It is issued against a patent generally immediately on expiry of the term of the patent. It cannot be issued before 5 years from filing date of corresponding patent.
7. The registration is required which is issued on national basis. A separately application is to be filed for registration in each country.

7. Indigenous Property Rights

The Indigenous Knowledge also known as Traditional Knowledge [TK] refers to the knowledge, innovations and practices of indigenous people and local communities. In other words, it is the: knowledge gained through tradition from ancestors about plants and animals, medicines, agricultural practices and other matters. Main points about traditional knowledge are listed below:

1. Traditional knowledge can be protected by sui-generic rights.
2. Traditional knowledge is also known as indigenous knowledge or local knowledge or traditional ecological knowledge.
3. It covers the matured long-standing traditions and practices of certain regional, indigenous, or local communities. In other words, it is gained from experience gained over the centuries and adapted to local culture and environment.
4. It includes knowledge about plants, animals, indigenous treatment of diseases through herbal medicines, cultural practices,etc.
5. Traditional knowledge is based on teachings and experiences and is orally passed on from generation to generation to some persons only. Earlier there were no written document on traditional knowledge.
6. Some forms of traditional knowledge are expressed through stories legends, folklore, rituals, songs and even laws.
7. There is difference between local knowledge and traditional knowledge The former existed from decade to century, whereas the latter existed from millennia.
8. It is collectively owned by the people or natives of an ecosystem or region

9. Traditional knowledge incorporates knowledge of ecosystem and code of ethics governing appropriate use of environment. This code includes rules and conventions promoting desirable ecosystem relations, human-animal interactions and even social relationships. A combination of traditional and non-traditional knowledge will form a rich and distinctive understanding of life and the world.

The ecological knowledge is composed of both traditional knowledge and experiential knowledge[gained through personal experience.

Traditional knowledge is very important almost in every sphere of life. Hence, its protection is very necessary otherwise this valuable knowledge will be gradually lost.

Summary

Sui-generis is a Latin expression, literally meaning of its own kind/genus or unique in its characteristics. In law, it is term of art used to identify a legal classification that exists independently of other categorizations because of its uniqueness or due to the specific creation of an entitlement or obligation. In Intellectual Property there are rights which are known as being sui-generic to owners of a small class of works. Such Intellectual Property Rights include (i) database Rights, (ii) mask work, (iii) moral rights, (iv) supplementary protection certificate (v) plant breeders rights and (vii) traditional knowledge, important points related to all above items have been discussed.

Glossary

Sui-generis Rights: Sui-generis refers to things of their own kind or things with unique characteristics.

Database Rights: A 'database right' is an intellectual property right given to a computer database.

Mask Work: A mask work is a two or three-dimensional layout of an integrated circuit

Moral Rights: Moral rights are a special extension to copyright that gives the originator of a copyrighted work rights over its use.

Supplementary Protection Certificate: This term is used in European Countries. This is an extension of the term given to pharmaceutical or plant protection patent.

Traditional Knowledge: The Traditional Knowledge [TK] refers to the knowledge, innovations and practices of indigenous people and local communities.

Plant Breeders' Rights: Plant breeders' rights, also known as plant variety rights (PVR), are intellectual property rights granted to the breeder of a new variety of plant.

Farmers' Rights: The legal rights provided to farmers to save, use, sow, replant, exchange, share or sell his farm produce including seed of a variety protected under Plant Variety Protection Act.

Questions

1. **What is sui-generic right ? List various items covered by such rights and describe any one of them in detail.**

2. **Define database and describe it significance.**

3. **Explain briefly disadvantages of sui-generic rights**

4. **Write short notes on the following.**

   a. Database rights
   b. Moral rights
   c. Mask work
   d. Traditional knowledge

5. **Define mask work and discuss its main features.**

6. **What do you mean by plant breeders' rights? Give merits and demerits of PBR.**

7. **Define traditional knowledge. Give its advantages and disadvantages.**

8. **Give a brief account of supplementary protection certificate.**

9. **What do the following terms signify?**

   a. Sui-generic rights
   b. Plant breeders' rights
   c. Farmers' rights
   c. Biodiversity

10. **Give a brief comparison of moral rights and traditional knowledge.**

11. **Describe briefly Farmers' Rights and their significance.**

Chapter 14

Database Rights

Introduction

The database refers to a collection of data organized for rapid search and retrieval by a computer. In other words, database refers to an organized collection of information, data, or citations stored in electronic format that can be searched for specific information or records by techniques specific to each database. A 'database right' is an intellectual property right given to a computer database. A database is defined as a collection of independent works, data or other materials which: (a) are arranged in a systematic or methodical way; and (b) are individually accessible by electronic or other means.

The database rights are of recent origin. The database rights were first introduced by European Union Law 1996. In U.K. it was introduced in 1997 but came into force from 1st January 1998. The purpose of this new database right is to protect the investment in a database. A database right arises automatically at no cost to you Main points about database rights are as follows:

1. Database rights were first adopted by U.K. in 1998.
2. Database rights are covered by sui-generic rights.
3. In most countries database rights are covered by copyrights.
4. Database rights exist in most European countries but not in the rest of the world.
5. Members of WTO have an obligation to provide copyright protection for some databases.
6. A database may be protected by copyright and/or the database right.
7. The protected database is permitted for normal usage [see later].

Main Features

The main features of related rights in relation to registration, requirements, matters covered, duration of protection, rights covered, authorization, enforcement, infringement etc. are briefly discussed as follows.

1. **Registration:** In many respects, a database right is similar to copyright. It is created automatically. In other words, this is an automatic right which protects the investment (*i.e.* time, money and energy) that goes into obtaining, verifying or presenting the contents of a database. There is no need of registration of database. But it is advisable to register the databse, because registration provides legal rights against unauthorised use of database by other persons. It is a right gainst use an dnot a monopoly.
2. **Requirements:** The owner of a protected database is a person who has taken the risk of investing in the creation of database. For the database rights to apply, the database must be author's own intellectual creation and the result of substantial investment (financial, technical or otherwise). A database will qualify for protection if it meets the following requirements:
   (i) **Author's own Creation:** The database is a *"collection of individual works, data or other materials arranged in a systematic or methodical way"*, and
   (ii) **A substantial investment:** There should be a proof that substantial investment has been made in creation and maintenance of database.
3. **Matters Covered:** Any work that shows originality in its selection, coordination and arrangement can be protected under database rights The following items are covered by database rights.
   1. Telephone directories or telephone book,
   2. Railway time tables,
   3. Electronic photo book,
   4. Encyclopedia with different entries,
   5. Microfilm collections, etc.

   The following items are not covered by database rights

   (i) Any computer software used to access or maintain the database.

   (ii) A movie, a recording, an audiovisual, cinematographic, literaryor musical work,
4. **Duration of Protection:** The term for protection of database rights is 15 years from the end of the year in which the database was completed or first made avilable to the public.The term is extended automatically, if a substantial change [additions, deletions or alterations] is made to the datbase during the term of protection The changes include substantial alterations, deletions and additions to the database in such a way so as to reflect substantial new investment.
5. **Rights Provided:** A database right prevents copying of substantial parts of a database (including frequent extraction of insubstantial parts). However unlike copyright the cover is not over the form of expression of information but of the information itself. The database owner is entitled to prevent others from:

(i) Extraction and or re-utilization of the whole or of a substantial part, evaluated qualitatively and/or quantitatively, of the contents of that database'.

(ii) Repeated and systematic extraction and/or re-utilization of insubstantial parts of the contents of the database implying acts which conflict with a normal exploitation of that database or which unreasonably prejudice the legitimate interests of the maker of the database'.

6. **Authorization:** The owner of the database has the rights to authorize other person to use his database on the basis of royalty payment. In other words, the holder of database has right to sale or licence his database.
7. **Enforcement:** The database rights come into force from the end of the year in which the the making of database was completed or the database was made available to the public.
8. **Infringement:** The rights holders may take action if their database is copied or used without permission. Database right can be infringed by:
   (i) Copying the contents of a database and rearranging them.
   (ii) Making the contents of the database available to the public without permission.
   (iii) Unauthorized repeated and systematic extraction of 'insubstantial parts' of database.
   (iv) Permanent or temporary transfer of the database contents to another medium by any means, or in any form.

As with copyright, if a database right is infringed a range of civil remedies are available to the database right holder including injunctions, an account of profits, damages and delivering up of goods.

Normal Usage of Database

The extraction or reutilization of substantial part of a published data is permitted without permission of the database creator under following circumstances:

(i) If the data is used for teaching or research and not for any commercial purpose.
(ii) When the identity of the data creator is not ascertainable or
(iii) When the term of database right is expired.

Database Rights and Copyrights

Database right is in many ways very similar to copyright so that, for example, there is no registration for database right – it is an automatic right and commences as soon as the material that can be protected exists in a recorded form. Ownership of database rights is determined in the same way as copyright ownership. As with copyright, database rights can be sold or licensed. The database right can apply to both paper and electronic databases. However, there are some differences between database right and copyright:

1. The term of protection for database right is much shorter. Database right lasts for 15 years from publication. It is possible that the right will last in perpetuity as the 15 year period of protection rolls forward if substantial changes and updates or other sufficient investments are made to the database in this period.
2. The activities that a database right holder can control, and which are, therefore, infringed if undertaken without the right holder's permission, are a bit different. Database right concerns control over the extraction and re-utilization of the contents of the database.

Table 14.1: Comparison of Database and Copyrights

Sl.No.	*Particulars*	*Database Rights*	Copyrights
1.	Governed by	Sui-generic rights	Primary intellectual rights
2.	Registration	Not compulsory	Not compulsory
3	Period of protection	15 years	70-120 years
4.	Protection over	Information	Expression
5.	Validity	Country of registration	Country of registration
6.	Enforcement	After end of the year of creation	After creation

Advantages of Database Rights

The Database Management System [DBMS] is a central system which provides a common interface between the data and the various front-end programs in the application. It also provides a central location for the whole data in the application to reside. The DBMS has several advantages over file based database management system as discussed below:

1. **Minimal Data Redundancy:** Since the whole data resides in one central database, the various programs in the application can access data in different data files. Hence data present in one file need not be duplicated in another. This reduces data redundancy. However, this does not mean all redundancy can be eliminated. There could be business or technical reasons for having some amount of redundancy. Any such redundancy should be carefully controlled and the DBMS should be aware of it.
2. **Data Consistency:** Reduced data redundancy leads to better data consistency.
3. **Data Integration:** Since related data is stored in one single database enforcing data integrity is much easier. Moreover, the functions in the DBMS can be used to enforce the integrity rules with minimum programming in the application programs.

4. **Data Sharing:** Related data can be shared across programs since the data is stored in a centralized manner. Even new applications can be developed to operate against the same data.
5. **Enforcement of Standards:** Enforcing standards in the organization and structure of data files is required and also easy in a Database System, since it is one single set of programs which is always interacting with the data files.
6. **Application Development Ease:** The application programmer need not build the functions for handling issues like concurrent access, security, data integrity, etc. The programmer only needs to implement the application business rules. This brings in application development ease. Adding additional functional modules is also easier than in file-based systems.
7. **Better Controls:** Better controls can be achieved due to the centralized nature of the system.
8. **Data Independence:** Data Independence can be defined as immunity of applications to change in physical representation and access technique. The provision of data independence is a major objective for database systems.

The architecture of the DBMS can be viewed as a 3-level system comprising the following:

(i) The internal or the physical level where the data resides.

(ii) The conceptual level which is the level of the DBMS functions.

(iii) The external level which is the level of the application programs or the end user.

Reduced Maintenance: Maintenance is less and easy, again, due to the centralized nature of the system.

Disadvantages of Database Rights

1. Database systems are complex, difficult, and time-consuming to design Substantial hardware and software start-up costs.
2. Damage to database affects virtually all applications programs.
3. Extensive conversion costs in moving form a file-based system to a database system.
4. Initial training required for all programmers and users.

Summary

The database refers to a collection of data organized for rapid search and retrieval by a computer. In other words, database refers to an organized collection of information, data, or citations stored in electronic format that can be searched for specific information or records by techniques specific to each database. A 'database right' is an intellectual property right given to a computer database.

The database rights were first adopted in UK in 1998. The database right allows the developers of a database to protect their database for up to fifteen years. The rights holders may take action if their database is copied or used without permission.

The main features of related rights in relation to registration, requirements, matters covered, duration of protection, rights covered, authorization, enforcement, infringement have been briefly discussed.

The comparison of database rights and copyriths is given in a tabular form. The advantages and disadvantages of database have been presented.

Glossary

Data: This term is legally defined according to the institution. It generally refers to recorded information regardless of form. Most institutions hold title to data while researchers have rights to access the data.

Database: A collection of data organized for rapid search and retrieval by a computer.

Database Rights: An intellectual property right given to a computer database.

Data Management: The process of handling the database by central database management system. May also refer to the department responsible for managing data entry and database generation and/or maintenance.

Data Definition Language (DDL): The language used to define objects in a database: CREATE TABLE, CREATE INDEX, and so on. DDL provides the semantics for administering all the physical objects in your database. It is based on a given standard, but may deviate some from vendor to vendor. All of the objects created together, form the 'schema' of a given application.

Data Manipulation Language (DML): The language used to manipulate objects in a relational database: We only have four statements in DML: INSERT, SELECT UPDATE and DELETE. DML provides the semantics for manipulating one or more occurrences (record) in a table in our database. It is based on a given standard, but may deviate some from vendor to vendor, especially through 'extensions' found valuable by the given vendor.

Data model: A data model is **not** the model (conceptual/logical/physical) of you future or existing database: It is a generic model which you base your analysis and design upon.

Database model: A Database model is the logical conversion of an Entity Relationship model. While an ER model reflects the business case, the Database model visualizes the logical model, which in turn is the basis for the physical implementation in the form of database tables, indexes and other mechanism in the database, necessary to construct the database in question.

Database Normalization: Database Normalization is the process of analyzing you database model to ensure that information is stored only in one place in th database, and that there are no derivates of the information within the database

Data-type: Identifies the kind of information that an Attribute/column in an entity/table on a specific database platform represents. These are actual physical representations and are dependent on the actual RDBMS's. The data-type should be given by the domain definition for that attribute/column.

Extraction: The permanent or temporary transferof all or a substantial part of the contents of a database to another medium by any means or in any form.

Re-utilization: Any form of making available to the public all or a substantial part of the contents of a database by distributing of copies, by renting, by on-line or other forms of transmission.

RDBMS: Short for Relational Database Management System and pronounced as separate letters, a type of database management system (DBMS) that stores data in the form of related tables. Relational databases are powerful because they require few assumptions about how data is related or how it will be extracted from the database. As a result, the same database can be viewed in many different ways.

MetaData: 'Data about Data'. This is the documentation stored in the database repository, and which holds information about your database objects. In Oracle, for example, the table USER_TABLES holds vital information about your tables.

NULL values: The concept of NULL, which means "unknown value", (not the digit zero) is not a part of the relational model, but has been introduced by the different vendors. It can create unexpected problems when accessing the database (read the linked article).

Occurrence: Each row in a database table. If you have 100 customers in your CUSTOMERS table, the table has 100 occurrences.

Open Database Connectivity (ODBC): A general interface for communication with different vendor-specific Relational Database Systems.

Questions

1. **Define database rights and describe their significance.**
2. **Describe briefly the advantages and disadvantages of database rights**
3. **Describe database rights on the basis of following points.**
   a. Registration
   b. Period of protection
   c. Rights covered
   d. Matters covered
4. **Describe briefly the matters that are covered by database rights.**
5. **What are the differences between database rights and copyrights? Describe in tabular form.**
6. **Explain briefly the main features of database rights.**

7. **Write short notes on the following:**
   a. Database rights
   b. Infringement
   c. Copyrights
   d. Meta-data

8. **List the matters covered by database rights. Describe their period of protection.**

9. **Explain briefly the following:**
   a. Matters protected under database rights
   b. Matters not protected under database rights

10. **What legal actions can be taken against infringement of database rights ?.**

11. **What are the normal usages of database?**

12. **Write short notes on the following**
    a. Extraction of protected database
    b. Re-utilization of protected database

Chapter 15

Mask Work Rights

Introduction

A **mask work** is a two or three-dimensional layout or topography of an integrated circuit (IC or "chip"), *i.e.* the arrangement on a chip of semiconductor devices such as transistors and passive electronic components such as resistors and interconnections. The layout is called a *mask* work because, in photolithographic processes, the multiple etched layers within actual ICs are each created using a mask, called the photomask, to permit or block the light at specific locations.

Mask work exclusive rights were first granted in the US by the Semiconductor Chip Protection Act [SCPA] of 1984. In Canada these rights are protected under the Integrated Circuit Topography Act 1990. Equivalent legislation exists in Australia and Hong Kong. Important points related to mask works are given below:

1. Mask work protection is protected by **sui-generis** rights.
2. The correct symbol used in a mask work's copyright notice is $ (**M** enclosed in a circle;
3. Mask work rights are closer to copyrights than with other exclusive rights such as patents or trademarks.
4. They are used alongside copyright to protect a read-only memory (ROM) component that is encoded to contain computer software.

Main Features

The main features of related rights in relation to registration, matters covered, duration of protection, rights covered, requirements, authorization, enforcement and infringement are briefly discussed as follows:

1. **Registration:** The registration is not compulsory for mask works. But registration is essential to get legal rights.The registration process consists of the following steps:

Filing Application: The application has to be filed in the Patents and Copyrights office in the prescribed form which cab be obtained from the copyrights office. The application for registration may be submitted by the owner of the mask work or an agent duly authorized by the owner.

Fee: The prescribed processing fee has to be deposited alongwith application form.

Declaration Certificate: A declaration has to be submitted stating that the applicant is authorized to submit the application and the statements made are correct to the best of knowledge of the person filing the application.

The application should name and signature [original] of the applicant.

For the same version of a mask work only one registration can be made.

Mask works must be registered within 2 years of first commercial exploitation. The date of registration is the date on which a complete application, fee, and an appropriate deposit representing an acceptable filing are filed with the Copyright Office. If the Mask Work is not registered within this time period, protection under the SCPA will be lost.

2. **Matters Covered:** Mask work rights cover two or three-dimensional layout or topography of an integrated circuit (IC or "chip"), *i.e.* the arrangement on a chip of semiconductor devices such as transistors and passive electronic components such as resistors and interconnections.
3. **Duration of Protection:** The rights in semiconductor mask work last only two years (if unregistered) or ten years (if registered). The SCPA provides protection of a Mask Work for a period of 10 years from the date of first commercial exploitation, or the date of registration with the Copyright Office, whichever occurs first.
4. **Rights Covered:** During the term of protection, the owner of a mask work has the following exclusive rights:
   (i) To reproduce the mask work by optical, electronic, or any other means
   (ii) To import or distribute a semiconductor chip product in which the mask work is embodied;
   (iii) To induce or knowingly to cause another person to do any of the acts described in number 1 and number 2.

   The exclusive rights in a mask work are somewhat like those of copyright the right to reproduce the mask work or (initially) distribute an IC made using the mask work. A owner of a mask work may freely import distribute or use, but not reproduce the chip (or the mask).
5. **Requirements:** The legal protection is provided to the three-dimensional images or patterns formed on or in the layers of metallic, insulating, or semiconductor material and fixed in a semiconductor chip product, *i.e* the "topography" of the "chip." For protection, a mask work must be original. A mask work is considered "original" if it is the independent creation of an author who did not copy it.

6. **Authorization/Transfer:** The Act allows the owner of the exclusive rights to transfer all of them or to license all or fewer than all the rights. A transfer or license must be in writing and signed by the owner of the rights or by a duly authorized agent of the owner. The Act also provides that the exclusive rights in a mask work may be transferred by operation of law or by terms of a will; or, the rights may pass as personal property under the applicable state Law. In other words, the owner of amask work can transfer all the exclusive rights of his mask work to any other person.The new owner should obtained by transfer the totality of the rights in the mask work
7. **Enforcement:** Protection for a mask work begins on the date the mask work is registered with the Copyright Office, or the date the mask work is first commercially exploited anywhere in the world, whichever occurs first. Protection lasts for 10 years.The protection expires at the end of the tenth calendar year after it began.

   The mask work rights come into force immediately after creation of mask work. In other words, mask work rights exist automatically [like copyrights] when they are created, regardless of registration. In case of patents, rights are confered only after registration.
8. **Infringement:** After a certificate has been issued by the Copyright Office, the owner of the mask work or the exclusive licensee of all rights in the mask work may institute a civil action for infringement occurring after the commencement of protection accorded by the Act.
9. **Notice and Symbol:** An appropriate notice on a semiconductor chip or product includes either the words "Mask Work," the symbol *M*, or the symbol Ⓜ, followed by the name of the owner or a recognizable abbreviation of the owner's name. Such notice is optional, but provides certain benefits to the owner with respect to enforcing rights.

Activities Permitted

It is important to understand that the Semiconductor Chip Protection Act [SCPA] does not provide as broad protection as the patent laws do. Certain activities by competitors are permitted, such as reverse engineering for the purposes of teaching, analyzing, or evaluating the concepts embodied in the mask work including the circuitry, logic flow, and organization of components. The results of legitimate reverse engineering may be incorporated into another original mask work and commercially exploited.

Any idea or concept, process, principle, discovery or system associated with a semiconductor product] a mask work is not protected.

Mask Work and Database Rights

There are some similarity and some differences between mask work and database rights which are presented in Table 15.1.

Table 15.1: Comparison of Mask Work and Database Rights

Sl.No.	Particulars	Mask Work	Database Rights
1.	Governed by	Sui-generic rights	Sui-generic rights
2.	Registration	Not compulsory	Not compulsory
3.	Period of protection	Unregistered 2 years Registered 10 years	15 years
4.	Protection over	Integrated Circuits	Information
5.	Validity	Country of registration	Country of registration
6.	Enforcement	After creation	After the end of the year of completion or creation

Advantages of Mark Work Rights

1. Registration is optional.
2. It protects original work.
3. It promotes creative work.
4. It provides benefits to the owner.

Disadvantages of Mass Work Rights

1. Protection is for shorter period than copyrights.
2. It does not provide legal rights without registration.
3. Rights are valid only in the country where registered.
4. It may lead to monopoly.

Summary

A **mask work** is a two or three-dimensional layout or topography of an integrated circuit (IC or "chip"), *i.e.* the arrangement on a chip of semiconductor devices such as transistors and passive electronic components such as resistors and interconnections.

Mask work exclusive rights were first granted in the US by the Semiconductor Chip Protection Act of 1984. In Canada these rights are protected under the Integrated Circuit Topography Act 1990. Equivalent legislation exists in Australia and Hong Kong.

The main features of related rights in relation to registration, matters covered, duration of protection, rights covered, requirements, authorization, enforcement and infringement have been discussed.

Comparison of copyrights and related rights has been presented in tabular form. Advantages and disadvantages of related rights have been outlined.

Glossary

Integrated Circuits: The final or intermediate form of any product (1) having two or more layers of metallic, insulating, or semiconductor material, deposited or otherwise placed on or etched away or otherwise removed from a piece of semiconductor material in accordance with a predetermined pattern; and (2) intended to perform electronic circuitry functions.

Mask Work: The layout is called a mask work because, in photolithographic processes, the multiple etched layers within actual ICs are each created using a mask, called the photo mask, to permit or block the light at specific locations, sometimes for hundreds of chips on a wafer simultaneously.

Mask Work Rights: The intellectual property right conferring time-limited exclusivity to reproduction of a particular layout.

Questions

1. **What do you mean by mask work rights? What is their significance?**
2. **Describe briefly the advantages and disadvantages of mask work rights.**
3. **Describe mask work rights on the basis of following points.**
   a. Matters protected
   b. Registration
   c. Rights covered
   d. Authorization
4. **Describe briefly the main features of mask work rights.**
5. **What are the differences between mask work rights and moral rights? Describe in tabular form.**
6. **Write short notes on the following:**
   a. Mask work rights
   b. Infringement
   c. Copyrights
   d. Patent rights
7. **Describe briefly the rights covered for the mask work.**
8. **Explain briefly the following:**
   a. Matters protected under mask work rights
   b. Relation of mask work rights with copyrights
9. **What legal actions can be taken against infringement of mask work rights?**
10. **Define the following terms:**
   a. Exclusive rights
   b. Mask work rights
   c. Copyrights
   d. Patent rights

Chapter 16

Plant Breeders' Rights

Introduction

The Plant Breeders' Rights programme was first established in 1987 under the Plant Variety Rights Act [PVRA], which was succeeded by the current Plant Breeders' Rights Act 1994. {PBRA].Plant breeders' rights [PBR], also known as plant variety rights (PVR), are intellectual property rights granted to the breeder of a new variety of plant. Plant Breeders Rights are granted to novel plant varieties that are distinctive, uniform, and stable (e. g, cultivars breed true-to-type for desired traits). The legal protection of a new plant variety is granted to the breeder or his successor. The effect of PBR is that prior authorization is required before the material can be used for commercial purposes.

Main Features

Main features of Plant Breeders' Rights in relation to registration, duration, validity, matters covered, requirements, transfer, control, enforcement, etc. are briefly presented below:

1. **Registration:** Registration is essential to get legal rights. There is no protection without registration. The registration consists of following steps:
   - (i) Filing application in the prescribed form,
   - (ii) Payment of prescribed processing fee along with application. The registration fee varies from country to country. In USA, after registration annual renewal fee [300 Dollars] is payable.
   - (iii) Examination of the application by the experts, and
   - (iv) Issue of protection certificate, if application if found to meet the desired requirements.
   - (v) Normally the registration takes about 2.5 years for most of the species. takes 2.5 years.

2. **Duration:** The period of protection varies with plant species. The protection is provided for a definite period. The duration of protection for different plant species is as follows:
   (i) For Trees and Vines: Initially for nine years and maximum for 18 years.
   (ii) For extant varieties: Initially six years and maximum for 15 years fro the date registration of the variety.
   (iii) In other cases: Initially six years and maximum for 15 years fro the date registration of the variety.

   Thus breeders' rights allow plant breeders to get benefit of their variety for a limited period. When the PBR period is over, the variety reverts to the public domain and is available to every body. The limited duration of Plant Breeders' Rights ensure a balance between private and public interest.
3. **Validity:** The protection of the Plant Breeders' Rights is valid only in the country where it has been registered. The protection in other countries can be obtained by filling separate application in each country.
4. **Matters Covered:** The protection right can be granted for varieties of all botanical genera and species. The variety should have a designation [name] as per the rule of International Code of Nomenclature.
5. **Requirements:** There are four basic requirements for protection of a variety under PBR, *viz.* (i) novelty, (ii) distinctiveness, (iii) uniformity, and (iv) stability. See later.
6. **Transfer:** An authorized plant breeder has right to authorize other interested persons for commercial production and marketing of his variety on royalty basis.
7. **Control:** The authorized plant breeder has right to prevent other from commercial production and marketing of his variety without permission.
8. **Enforcement:** The Plant Breeders' Rights come into force immediately after registration of the variety.
9. **Other Points:** There are some other points which are taken into account for granting PBR. Thesr are as follows:
   (i) BPR can be granted only to breeder or legal owner of a new variety
   (ii) The plant breeders rights protect the variety but not the standard breeding procedures that are used to develop new variety.
   (iii) The main purpose of PBR is to encourage breeders to develop better plant varieties.
   (iv) The PBRs provide monopoly of new varieties to breeders for a limited period.
   (v) The same variety can be patented and protected by PBR. The gene technology or gene sequences can be patented and the end product that is the plant variety can be protected under PBR.

(vi) A person growing protected variety can not sale its seed on commercial scale. But he can use the produce of such seed for planting purpose on his farm.

(vii) PBR coexists with other legislation. It is not independent of other laws of the land.

(viii) The protected variety can be used to breed new varieties.

(ix) A variety cannot be registered under trademarks. It comes under Sui-generic rights, hence it can be registered under PBR only.

Requirements

The PBR cannot be granted to any variety. There are four basic requirements for protection of a variety under Plant Breeders' Rights. These are (i) novelty, (ii) distinctiveness, (iii) uniformity, and (iv) stability. These are defined as follows:

1. **Novelty:** It refers to newness of a variety. The variety should be new and should not have been commercially cultivated for more than one year before granting protection under PBR. In other words, there should not be a prior sale for more than one year.
2. **Distinctiveness:** The new variety must be distinguishable in one or more morphological, quality or other character from previously available varieties. When a variety differs atleast in one characteristic from varieties of common knowledge, it is known as distinct.
3. **Uniformity:** The variety should be uniform. In other words, all plants in a variety should look alike.
4. **Stability:** The should give stable performance in different generations. All these attributes are determined by grow out test.

Rights Protected

The *Plant Breeders' Rights Act* provides plant breeders the exclusive right to produce and sell new plant varieties developed by them. In other words, It provides exclusive rights to the breeder for commercial production and marketing of his variety. The important Plant Breeders' Rights are as follows:

1. **Rights for commercial Seed production:** PBR provide legal right to the breeder of a variety for large scale seed production. This he can do either on his own farm or on the farms of authorized farmers on payment basis.
2. **Rights for Marketing:** The breeder or owner of a variety has exclusive rights to regulate marketing of his variety.
3. **Rights to export and import:** The breeder or owner of a variety has full rights to regulate export and import of his or her variety.
4. **Rights of Authorization:** The breeder or owner of a protected variety has rights to authorize other interested persons for commercial production, and marketing, export and import of his variety. However, prior authorization of the breeder or owner of a variety is required for such purpose.

5. **Rights to prevent Infringement:** The breeder or owner of a variety has rights to prevent others from unauthorized commercial production and marketing of his variety.

Infringement

Unauthorized production and marketing of a registered variety by other person amounts to infringement. The owner has the right to take legal action against the infringer and claim damages. The PBR Authority may initiate legal action against a person who is involved in the infringement of a protected plant variety. The Authority can recover both damages and profits from such person. The Plant Breeders' Act [PBRA] provides for heavy penalties against infringement of the breeders' right. In USA it 55,00 Dollars for individuals and 275000 Dollars for companies.

Exemptions of PBRA

The Plant Breeders' Rights Act provides two exemptions, *viz.* (i) breeders' exemptions and (ii) farmers' exemptions. These are defined as follows:

(i) **Breeders' Exemptions:** The legal rights that are provided to plant breeders to use protected material for further research refer to breeders' exemptions. Breeders exemptions are also called research exemptions or breeders' privilege. The UPOV Act 1978 provides Breeders' exemptions. However, the Act 1991 has curtailed Breeders' Exemptions.

(ii) **Farmers' Exemptions:** The legal rights that are provided to farmers to save, use, exchange, share or sell his farm produce of a protected variety are known as farmers' exemptions. Farmers' exemptions are also called Farmers' Rights or Farmers privilege. Here the sale is restricted to non-commercial sale. The UPOV Act 1978 provided farmers Exemptions however, these exemptions were curtailed by UPOV Act 1991.Because of these reasons, UPOV Act is not accepted by many countries.

Advantages of PBR

1. Breeders get benefit of their variety.
2. PBR help in faster development of seed industry.
3. PBR lead to improvement in quality because of competition.
4. PBR are useful in procurement of good material on payment basis.
5. PBR help in enrichment of genetic resources.

Disadvantages of PBR

1. It will promote monopoly.
2. It will encourage unhealthy practices.
3. It may lead to increase in prices.

4. There will be reduction in genetic variability.
5. There will be compulsion to purchase fresh seed every year.

Summary

The Plant Breeders' Rights programme was first established in 1987 under the Plant Variety Rights Act [PVRA], which was succeeded by the current Plant Breeders' Rights Act 1994. {PBRA].Plant breeders' rights [PBR], also known as plant variety rights (PVR), are intellectual property rights granted to the breeder of a new variety of plant. Plant Breeders Rights are granted to novel plant varieties that are distinctive, uniform, and stable (e.g, cultivars breed true-to-type for desired traits). The legal protection of a new plant variety is granted to the breeder or his successor. The effect of PBR is that prior authorization is required before the material can be used for commercial purposes.

Main features of Plant Breeders' Rights in relation to registration, duration, validity, matters covered, requirements, transfer, control, enforcement, etc. have been described.

There are four basic requirements for protection of a variety under Plant Breeders' Rights. These are (i) novelty, (ii) distinctiveness, (iii) uniformity, and (iv) stability.

The plant breeders' rights Act provides (i) right to commercial seed production, (ii) right to marketing, (iii) right export and import (iv) right of authorization, and (v) right to prevent infringement for his variety.

The Plant Breeders' Rights Act provides two exemptions, *viz.* (i) breeders' exemptions and (ii) farmers' exemptions. The advantages and disadvantages of Plant Breeders' Rights Act have been outlined.

Glossary

Breeders' Exemption: Rights provided to use protected material as the basis to develop a new variety or for other research work; also called research exemptions or breeders' privilege.

Distinctiveness: The variety must be distinguishable in atleast one character from previously available varieties.

Farmers' Exemption: Legal rights provided to farmers to save, use, exchange, share or sell his farm produce of a protected variety; also called farmers' rights or farmers' privilege.

Novelty: The newness of a variety means not previously sold.

Plant Breeders' Rights: Plant breeders' rights [PBR], also known as plant variety rights (PVR), are intellectual property rights granted to the breeder of a new variety of plant.

Stability: Each generation of the variety should be similar.

Uniformity: All plants in a variety should look alike.

Questions

1. **Define Plant Breeders' Rights and describe their main features.**
2. **What are advantages of Plant Breeders' Rights?**
3. **Describe briefly disadvantages of Plant Breeders' Rights.**
4. **Explain briefly the following:**
   (i) Breeders Exemption
   (ii) Farmers Exemption
5. **Write short notes on the following:**
   (i) Novelty
   (ii) Distinctiveness
   (iii) Uniformity
   (iv) Stability
6. **Discuss briefly the basic requirements for protection of plant variety under PBR.**
7. **Describe Plant Breeders' Rights on the basis of following:**
   (i) Registration
   (ii) Period of protection
   (iii) Rights covered
   (iv) Material covered
8. **Give the full forms of the following:**
   (i) PVR
   (ii) PBR
   (iii) IPR
   (iv) UPOV
9. **Define Infringement. What are the penalties for infringement of PBR ?**
10. **Discuss briefly the rights granted to a breeder for his variety under PBR.**
11. **What do the following terms signify?**
    (i) Breeders' privilege
    (ii) Farmers' privilege
    (iii) Novelty
    (iv) Stability

Chapter 17

Farmers' Rights

Introduction

Farmers' rights refer to the rights arising from the past, present and future contributions of farmers in conserving, improving and making available plant or animal genetic resources, particularly those in the centres of origin/diversity. In other words, the legal rights provided to farmers to save, use, sow, replant, exchange, share or sell his farm produce including seed of a variety protected under Plant Variety Protection Act refer to Farmers Rights. The purpose of these rights is to 'ensure full benefits to farmers and support the continuation of their contributions.

The FAO Conference held in Rome from 11-29 November, 1989 endorsed the concept of Farmers Rights with a view to:

(i) Ensuring global recognition of the need for conservation and the availability of sufficient funds for these purposes;

(ii) Assisting farmers and farming communities throughout the world, especially those in areas of original diversity of plant genetic resources, in the protection and conservation of their PGR and of the natural biosphere; and,

(iii) Allowing the full participation of farmers, their communities and countries in the benefits derived, at present and in the future, from the improved use of PGR.

Main Features

Main features of Farmers Rights in relation to registration, requirements, fee, rights covered, rights of communities, authorization, materials covered, duration, validity, benefit sharing, gene fund, rewards, infringement etc are briefly presented below:

1. **Registration:** A farmer who has bred or developed a new variety shall be entitled for registration and protection of his variety like a breeder. The registration process consists of the following steps:
   (i) **Filing of application:** The application is to be filed in the prescribed form and submitted in the office of the registrar of Plant Variety Protection and farmers' Rights Act. The application should contain all the desired documents and information.
   (ii) **Examination of the application:** The application is examined by experts in the registrar office. If found correct, it is advertised for opposition. Three months time is given for filing opposition.
   (iii) **Registration:** If there is no opposition, and the registrar and Authority are satisfied, the variety is registered and the registration certificate is issued to the breeder, farmer or other owner of the variety.
2. **Requirements:** The farmers; variety shall be entitled for registration if it meets the following requirements:
   (i) It should have a denomination [name] assigned by the applicant
   (ii) An affidavit that the variety does not contain terminator gene.
   (iii) A complete passport data of parental lines from which the variety has been derived.
   (iv) The contribution of any farmer, village community, institution or organization in the development of the variety.
   (v) The source of genetic material used in developing the variety and the parental material used in developing new variety was lawfully acquired.
   (vi) A brief description of the variety about its novelty, distinctiveness uniformity and stability.
   (vii) A proof of the right to make the application.
3. **Exemption from Fee:** A farmer or group of farmers or village community shall not be liable to pay any fee in any proceeding before the Authority or Registrar or the Tribunal or the High Court under Plant Variety Protection and Farmers' Rights Act or the rules made there under.
4. **Rights Covered:** The Farmers' Rights Act provide following rights to the farmers in India even for the variety protected under PVP and FR Act.
   (i) Rights to save, use, sow or replant seeds of his farm produce.
   (ii) Rights to exchange or share his farm produce, and
   (iii) Rights to sell seeds of his farm produce

   However, the farmer cannot sell the seeds or planting material under brand name.
5. **Rights of Communities:** In India, any of the following can file application for registration on behalf of any village or community.
   (i) Any person or group of persons [whether actively engaged in farming or not], and

(ii) Any governmental or non-governmental organization

After completing all formalities, if the Authority is satisfied, the variety can be registered. If due credit has not been given to the farmers or communities who have helped in the development of the variety, the breeder has to pay compensation to the concerned party. Or part of compensation will be deposited in the gene fund.

6. **Authorization of Farmers' Variety:** The authorization for an essentially derived variety from farmers' variety cannot be given by the breeder of such farmers; variety except with the consent of the farmers or group of farmers or community of farmers who have made contribution in the preservation or development of such variety.
7. **Materials Covered:** Farmers' Rights include new varieties of all self pollinated, cross pollinated and asexually propagated varieties.
8. **Duration:** The protection is provided for a definite period. The duration of protection for different plant species is as follows'
   (i) For Trees and Vines: Initially for nine years and maximum for 18 years.
   (ii) For extant varieties: Initially six years and maximum for 15 years fro the date registration of the variety.
   (iii) In other cases: Initially six years and maximum for 15 years fro the date registration of the variety.
9. **Validity:** The protection is valid only in the country where the variety has been registered.
10. **Benefit Sharing:** If the variety has been developed by a particular farmer, he can get benefits of his variety. In case of community rights, the sharing of benefits becomes difficult and the money is deposited in the gene fund.
11. **Gene Fund:** The central government may constitute a Fund to be called National Gene Fund. The money earned from following will be deposited in this fund;
    (i) The benefit sharing received from the breeder of a variety.
    (ii) The annual payable to the Authority in the form of royalty.
    (iii) The amount of compensation received by the Authority.
    (iv) Contribution received from amy National or International Organization and other sources.

    The gene fund may be utilized for benefit sharing, rewards and other expenditure related to Plant Variety Protection and Farmers' Rights Act.
12. **Rewards:** A farmer who is engaged in the conservation of genetic resources of land races and wild relatives of economic plants and their improvement through selection shall be entitled for recognition and reward. The expenses toward rewards will be met out from the National Gene Fund
13. **Acknowledgement by Breeders:** It is compulsory for the breeder of a variety to reveal the origin of the parental lines plant genetic resources used in the development of a variety. It will help in deciding the share of

different persons, *viz.*, farmers and other communities in benefits earned by breeder of such variety.

14. **Infringement:** If a farmer was not aware of the PVP and FR Act at the time of using any protected variety, he will be exempt from the infringement and will be asked not to do such act in future. This right has been provided to Indian farmers by PVP and FR Act 2001.

Nature and Scope of FR

Farmers' Rights, as endorsed by FAO in 1989, recognize that farmers and rural communities have contributed and continue to contribute to creation, conservation, exchange and enhancement of genetic resources and they should be recognized and strengthened in their work. The important points about nature and scope of Farmers' Rights are briefly presented as follows:

1. **Ancient Rights:** Farmers' Rights have a deep historic character. They existed since humans created agriculture to serve their needs, and played vital role in conservation of biodiversity. Farmers are the guardians of these genetic resources, which support the evolution of species. Farmers are the inheritors of the skills and knowledge of the generations that have created this biological wealth. For these contributions, it is necessary to recognize Farmers' Rights.
2. **Rights Covered:** Farmers' Rights include the right over resources and associated knowledge, united indivisibly, and mean the acceptance of traditional knowledge, respect for cultures and recognition that these are the basis of the creation of knowledge.
3. **Property Rights:** The farmers deserve right to control, the right to decide the future of genetic resources, the rights to define the legal framework of property rights of these resources.
4. **Collective Nature:** Farmers' Rights are of an eminently collective nature and therefore are protected by sui-generic rights.
5. **Application:** These rights should have a national application, respecting the sovereignty of each country, to establish local laws based on these principles.
6. **Rights to Means:** It includes territorial rights, right to land, right to water and air and rights to conserve biodiversity and achieve food security.
7. **Rights for Active Participation:** The farmers deserve right to participation in the definition, elaboration and execution of policies and programmes linked to genetic resources.
8. **Rights to Technology:** The farmers deserve right to appropriate technology as well as participation in the design and management of research programmes.
9. **Rights to Benefit Sharing:** The farmers deserve right to define the control and handling of benefits derived from the use, conservation and management of these resources.

10. **Rights on Genetic Resources:** The farmers deserve right to use, choose, store and freely exchange of genetic resources.
11. **Rights for Sustainable Agriculture:** The farmers deserveright to develop models of sustainable agriculture that protects the biodiversity and to influence the policies that support it.

Role of Farmers Related to PGR

Farmers are both men and women who have domesticated, developed, conserved and made available plant genetic and other natural resources (land; water; vegetation; animal, bird and fish life) to which they have access in order to obtain a livelihood and ensure the well-being of their family through the provision of basic requirements such as food, fuel and water and income from the land.

Farmers' Rights are not assigned to specific varieties, types of plants, or to specific farmers. Their purpose is to encourage farmers and farming communities to nurture and conserve, and to utilize and improve, plant genetic resources (FAO, 1996).

Farmers' Rights are generally considered to be group (collective) rights that do not necessarily imply direct compensation or bilateral reciprocity, such as selling a license for exclusive rights to collect and use crop germplasm.

Farmers have played key role in conservation of Plant Genetic Resource for thousands of years. The role of farmers in relation to PGR is briefly presented below:

1. Farmers both men and women have conserved plant genetic resources over the past thousands of years.
2. Farmers are rightly called as on site managers of plant genetic resources.
3. Farmers have in depth knowledge of the land races and wild relatives of cultivated plant species.
4. All modern plant varieties contain only those genes that have originated from farmers' traditional varieties (landraces) or wild crop relatives
5. The women and men have historically provided the germplasm upon which scientific plant breeding is based.
6. Generally, germplasm is provided by small-scale subsistence farmers to plant breeders whose varieties are then adapted/developed for medium to large scale commercial farmers (Spillane, 1996).
7. Women's special knowledge of the value and diverse uses of plants for nutrition, health and income has important implications for the conservation of plant genetic resources.
8. The women have played significant role not only in terms of labour and skills, but also their decision-making about how natural resources are used to satisfy the multiple needs of rural households.
9. In much of the developing world, the on-farm and in-situ conservation and use of plant genetic resources begins with women. As farmers, they

are responsible for growing crop plants and collecting seed for next year sowing.

10. It is estimated that ninety per cent of the planting materials used in developing countries is produced by the farmers themselves, *i.e.* from saved seed (Spillane, 1996). The selection of wild genetic resources for home planting, of seeds for conserving for next year's planting and of asexually propagated material is usually carried out by women. This selection is a sophisticated process that takes into account many different criteria such as taste, color, palatability and texture; resistance to pests and diseases; adaptation to soil and agro-climatic conditions and so on. The selection criteria are built up over years of experience.

PBR and Farmers' Rights

Plant Breeders' Rights and Farmers' Rights have some similarities and some difference. A brief comparison of these two is presented in Table 17.1.

Table 17.1: Comparison of Plant Breeders' Rights and Farmers Rights

Sl.No.	*Particulars*	*Plant Breeders' Rights*	*Farmers' Rights*
A.	**Similarities**		
1.	Registration	Essential	Essential
2.	Duration		
	Trees and vines	18 years	18 years
	Extant varieties etc.	15 years	15 years
3.	Enforcement	After Registration	After Registration
4.	Requirements	Novelty, Distinctiveness, uniformity and stability	Same as in PBR
5.	Validity	In country of registration	In country of registration
B.	**Dissimilarities**		
1.	Registration Fee	Charged	Not charged
2.	Annual renewal fee	Charged	Not charged
3.	Penalty for infringement	Charged	Not charged
4.	Community Rights	Not covered	Covered
5.	Authorization	Easy	Not permitted

Advantages of Farmers' Rights

1. Provide recognition to farmers, communities or village for their role in conserving Plant Genetic Resources.
2. Provide respect to farmers.
3. Provide opportunities to farmers to participate in the policies, projects and programmes related to Plant Genetic Resources [PGR].
4. Lead to benefit sharing when farmers material is used in developing new variety by breeders.

5. Farmers permission is required to use their plant material.
6. Provide rewards to those farmers who are actively engaged in conserving Plant Genetic Resources.

Limitations of Farmers' Rights

1. Mostly farmers' rights are of collective nature.
2. The benefit sharing is difficult.
3. Farmers do not get any financial support from government for the conservation of Plant Genetic Resources.
4. Farmers' Rights are of territorial nature.

Future Outlook Related to PGR

Plant Genetic Resources are the base materials for modern plant breeding programmes Hence their proper conservation and sustainable use are essential. In future, the following points should be given due importance for proper conservation and utilization of plant genetic resources.

1. **Assistance to Farmers:** Farmers in all regions of the world should be assisted in their conservation of natural resources and more specifically the selection, conservation, improvement and sustainable use of PGR. Non Governmental Organizations should assist in the entire process as they are close to the communities and have the capacities to provide long-term support. The establishment of strong village level institutions with adequate representation of both women and men farmers can be key in promoting successful programmes.
2. **Incentives:** Appropriate incentives should be developed, with the active participation of those concerned, to convince communities, industry and governments of the benefits of conservation and development efforts and to provide financial resources, legal support, and technical capability to develop agricultural systems that are suitable for local agro ecological conditions and, where possible, based on locally available resources. Incentives should be used to enable farmers to divert land, labour and capital towards conserving biological resources and to facilitate the participation of certain groups in activities that promote the sustainable use of those resources.
3. **Participation of Women Farmers:** The participation of women farmers and the recognition of their contributions in the design and elaboration of such policies and programmes at all levels is essential. Women farmers should be included in decision making processes at community and regional levels. Farmers, communities and countries, and men and women at all levels should participate in the design and implementation of programmes policies and agreements related to PGR.
4. **Self Decision:** Those farmers who develop breeding materials should be empowered to decide, at their discretion, if the materials will be made available to other (*i.e.* external) users.

5. **Equal Rights to Women:** There should be equal rights for conservation and participation in PGR related programmes and policies, which is often lacking in developing countries. There should not be discrimination of gender. In the past, women farmers could get little benefit from support in terms of information and technology, inputs and credit, extension and research.
6. **Participatory Plant Breeding:** Participatory research and plant breeding should be promoted with different socio-economic groups in rural communities in order to improve the productivity of local varieties and to promote marketing of special products derived from these materials.
7. **Use of Land Races:** Modern plant breeding has concentrated on developing a small number of varieties that are widely adapted to many environments. Land races should also be used for developing varieties suitable for diverse ecological conditions.
8. **International Funds:** There is need for the creation of an Intentional Fund to support in-situ conservation and to benefit rural men and women who produce and conserve PGR. In general farmers are interested in conserving resources but lack of financial support acts as a barrier in their way.
9. **Integrated Programs:** PGR conservation should be integrated into other community development programmes It will promote and employment in activities related to PGR building up of local institutions, capacity building) and collecting and dissemination;, information on the status and trends of PGR.
10. **Benefit Sharing:** There should be equitable sharing of benefits arising from the utilization of PGR, knowledge, innovations and practices.

Summary

The concept of Farmers' Rights was endorsed by the FAO Conference held in Rome from 11-29 November, 1989. **Farmers' rights refer to the** rights arising from the past, present and future contributions of farmers in conserving, improving and making available plant or animal genetic resources, particularly those in the centre of origin/diversity. In other words, the legal rights provided to farmers to save, use sow, resow, exchange, share or sell his farm produce including seed of a variety protected under Plant Variety Protection Act refer to Farmers Rights.

The purpose of these rights is to "ensure full benefits to farmers and support the continuation of their contributions.

Main features of Farmers Rights in relation to registration, requirements, fee rights covered, rights of communities, authorization, materials covered, duration validity, benefit sharing, gene fund, rewards, infringement etc have been discussed

The important points about nature and scope of Farmers' Rights have been explained. The role of farmers in relation to conservation of plant genetic resources has been amply discussed. The comparison of Plant Breeders' Rights and Farmers Rights has been given in tabular form. The advantages and limitations of Farmers Rights are listed and future outlook of plant genetic resources have been discussed

Glossary

Biological Diversity: Biological Diversity means the "variability among living organisms from all sources including, inter alia, terrestrial, marine and other aquatic ecosystems and the ecological complexes of which they are part; this includes diversity within species, between species and of ecosystems."

Biotechnology: Biotechnology means any technological application that uses biological systems, living organisms, or derivatives thereof, to make or modify products or processes for specific use.

Cultivar: Cultivated variety of plants.

Domestic biodiversity: The genetic variation existing among the species, breeds, cultivars and individuals of animal, plant and microbial species that have been domesticated, often including their immediate wild relatives.

***Ex-situ* conservation:** The conservation of components of biological diversity outside their natural habitats, often in a laboratory, collection, botanical garden, zoo, or aquarium.

Farmers' Rights: Rights arising from the past, present and future contributions of farmers in conserving, improving and making available plant or animal genetic resources.

Genetic resources: Genetic resources means genetic material of actual or potential value

***In-situ*:** In the original location. (Opposite: *ex-situ*).

***In-situ* conditions:** The conditions where genetic resources exist within ecosystems and natural habitats; in the case of domesticated or cultivated species, in the surroundings where they have developed their distinctive properties.

Landrace: A crop cultivar or animal breed that evolved with and has been genetically improved by traditional agriculturists, but has not been influenced by modern breeding practices.

Public Domain: Public Domain refers to the knowledge that is freely available, commonly shared throughout the world without any access restrictions. It cannot be protected by intellectual property rights.

Questions

1. **Define Farmers' Rights and describe their significance.**
2. **Describe briefly main features of Farmers' Rights.**
3. **Explain briefly the procedure of registration of Farmers' variety.**
4. **Describe Farmers' Rights on the basis of following:**

   (i) Registration (ii) Duration of protection

   (iii) Rights covered (iv) Authorization

5. **Write short notes on the following:**
   (i) *In-situ* conservation
   (ii) *Ex-situ* conservation
   (iii) Bio-diversity
   (iv) Land Race

6. **What do the following terms signify?**
   (i) Farmers' Rights
   (ii) Genetic Resources
   (iii) Public Domain
   (iv) Cultivar

7. **Give a brief comparison of Plant Breeders' Rights and farmers' Rights in a tabular form.**

8. **Describe the role of farmers in relation to Plant Genetic Resources.**

9. **Discuss the following in relation to Farmers' Rights:**
   (i) Infringement
   (ii) Registration Fee
   (iii) Role of women farmers
   (iv) Benefit sharing

10. **Describe the role of women farmers in relation to Plant Genetic Resources.**

11. **Describe briefly ways and means for sustainable use of Plant Genetic Resources.**

Chapter 18

Moral Rights

Introduction

Moral rights are a special extension to copyright that gives the originator of a copyrighted work rights over its use. The originator can specify moral conditions on the use or exploitation of their creation, even when the rights are licensed or sold to others'

An author is said to have the "moral right" to control his or her work. The concept of moral rights thus relies on the connection between an author and his/her creation. The main points related to moral rights are presented below:

1. These are rights provided to authors and directors of copyrights to protect their personal interest in relation to their creations. These rights are provided beyond those strictly recognized in copyright law.
2. Moral Rights were granted to authors under the Berne Convention (1989).
3. The creator of the work has the right of attribution and integrity, even after sale or transfer of the copyright. In other words, the creator or author shall have the right to claim authorship of the work even after sale or transfer of the copyright.
4. **Transfer:** The copyrights can be transferred or assigned to others, but moral Right cannot be transferred or assigned.
5. The author has right to object to any distortion, mutilation or other modification which would be prejudicial to his honor or reputation.
6. Moral Rights include the right of attribution, the right to have a work published anonymously or pseudonymously, and the right to the integrity of the work.
7. Moral Rights protect the personal reputation of the author/creator of copyright work.
8. Protection of moral rights ends with the death of the author.

Main Features

The main features of moral rights in relation to registration, matters covered, duration of protection, rights covered, requirements, authorization, enforcement and infringement are briefly discussed as follows.

1. **Registration:** Like copyrights registration is not compulsory for moral rights. But it is desirable to register moral rights to have legal control over the unauthorised use of moral rights.
2. **Matters Covered:** All those matters which are covered under copyrights are also covered by moral rights.
3. **Duration of Protection:** The term of moral rights is the same as the term of copyright protection. So when copyright protection ends and a work falls into the public domain the moral rights protection ceases at the same time. Moral rights are not transferrable, and end only with the life of the author, whereas copyrights continue up to 70 years after the death of the author.
4. **Rights Covered:** Moral rights are separate rights to that of copyright and belong to the author of a work. The following four categories of moral rights are associated with authorship, and are quite distinct from copyright [Copyright Designs and Patents Act 1988].
   (i) **Right to Paternity :** This right requires acknowledgment of the identity of the author. In other words, this right states that the authorship should be properly acknowledged.
   (ii) **Right to Integrity:** This right prevents derogatory treatment of authors' work. The right of integrity protects an author's work from intentional distortion, mutilation, or other modification that is likely to harm the author's honor and reputation, and the destruction of any work of recognized stature This right also protects authors from having their works associated with products, services, causes or institutions that would harm their honor or reputation.
   (ii) **Right to Attribution**: This right is not to have a work falsely attributed to an author. The right of attribution allows an author to prevent misattribution of his work. This right prevents the use of one's name on any work which the author did not create This right expires 20 years after the death of the author.
   (iv) **Right to Disclosure or Privacy**: This is the right to withhold publication of an image or broadcast. This requires that the authorship of the work should not be disclosed (*i.e.* remains anonymous) where it may tarnish his image. It is applicable in respect of certain photographs and films only.

The author has right to prevent others to alter his work without his permission to dictate in what way the work can be displayed and to receive resale royalties. The author can also prevent revision, alteration, or distortion of his work, regardless of who owns the work.

In Canada, copyright law provides the creators of works protected by copyright with several "moral" rights. These include:

1. **The right of association**: creators have legal recourse against unauthorized use of their work in association with products, services, causes or institutions in ways which are prejudicial to their honour or reputation;
2. The right to remain anonymous; [**Right of Non-disclosure**].
3. The right to use a pseudonym; [**Right of Non-disclosure**].
4. The right to be identified with the work as its author; [**Right of Paternity**].
5. The right of integrity: creators are protected against "distortions, mutilations or other modifications" to a work which are prejudicial to their honour or reputation. The law assumes that any changes to a unique work (*e.g.* a painting or sculpture) are prejudicial.

In United Kingdom, the Copyright, Designs and Patents Act 1988 came into force on August 1, 1989. The following five categories of moral rights are protected.

1. **Identification as author**: This gives creators the right to be identified as author, whenever the work is published commercially, exhibited in public, or included in a TV programme or film. The artists whose works are exhibited in public, especially sculptors and artists/craftspeople whose works are permanently sited in or on buildings and in spaces open to the public, can insist on their identification as author.
2. **No unjustified modification**: This right prevents unjustified modification of author's work whenever the work is published commercially, exhibited in public, or included in a TV programme or film. Modification means any addition, deletion, alteration or adaptation of the art work which prejudice the honour and reputation of the author.
3. **No false attribution as original author**: This right gives creators the right not to be named as authors of works which they did not create, whenever such works are shown to the public, through exhibition or publication of copies including film and TV programmes.
4. **No false attribution as author of altered work**:This right gives creatots the right not to be named as author in the work that has been without his consent. This also prevents anyone who deals with such work or copies of such work by way of trade.
5. **No false attribution as author of a copy of work:**This right enables artists to prevent dealers falsely trading in copies of their works purporting to have been authorised by them.

   Duration of Protection: All five rights will last for the author's lifetime and for a period after death. The first two, the right to be identified and not to suffer unjustified modification, will continue for 50 years beyond the author's death and will be capable of being left in artists' wills. The last three rights concerning false attributions, will continue for 20 years beyond the author's death but will not be capable of being left in an artist's will Exceptionally, in the case of photographs, all five rights will only last for 50 years from the making of the image.

Transfer: None of the moral rights will be capable of being transferred or assigned by the artist to anyone else during his/her lifetime.

5. **Requirements:** Since moral rights are linked with copyrights, there are no special requirements for protection of moral rights.
6. **Authorization/Transfer:** Moral rights related to copyright work are considered personal to the author and that cannot therefore be bought, sold or transferred. The moral rights cannot be assigned, but can be waived contractually. Under Canadian copyright law a creator cannot assign moral rights. It is only possible for a creator to agree not to exercise moral rights by "waiving" the rights. Authors may, however, waive their moral rights if do so in writing. Moral rights can be waived for one particular individual or for one particular use.
7. **Enforcement:** The moral rights become effective immediately after release of the copyrighted works.
8. **Infringement:** The determination of infringement involves two types of components, *viz.* a subjective and an objective component. The subjective component is whether the author himself feels that his work is used in a manner which is prejudicial to his honour or reputation. The objec component is an evaluation of public or expert opinion. The word of the author alone is not enough to establish infringement. To sustain a claim of infringement, an author must be supported by expert or public opinion. There are various mechanisms with which an author may enforce his moral rights. These include the following:
   - (i) An author may show that, in altering or distorting her work, someone has created a "derivative work," thereby violating the Copyright Act.
   - (ii) If authorship of a work is attributed to an author against her will or misattributed, the author may have a state action for defamation against the person responsible for the attribution.
   - (iii) If a person uses the identity of an author, or the works of the author for her own benefit without the author's permission, then she may have violated the author's right of publicity or may be guilty of misappropriation of the author's work.

   The person entitled to any of the moral rights can take legal proceedings in the civil courts for preventing infringements. He can ask for financial compensation for any damage caused by infringer including cost of legal proceedings.

Moral Rights and Mask Work Rights

There are some similarity and some differences between moral rights and mask work rights which are presented in Table 18.1

Table 18.1: Comparison of Moral Rights and Mask Work Rights

Sl.No.	*Particulars*	*Moral Rights*	*Mask Work*
1.	Governed by	Sui-generic rights	Sui-generic rights
2.	Registration	Not compulsory	Not compulsory
3.	Period of protection	Till the life of the author	Unregistered 2 years Registered 10 years
4.	Protection over	Copyrighted matters	Integrated Circuits
5.	Validity	Country of registration	Country of registration
6.	Enforcement	After release of copyrighted work	After creation
7.	Requirements	Copy righted material	Originality of mask work
8.	Transfer	Not permitted	Permitted
9.	Infringement	Possible	Possible

Advantages of Moral Rights

1. Registration is not compulsory.
2. Protection is life long.
3. It promotes creative work.
4. It provides benefits to the owner.

Disadvantages of Moral Rights

1. It is not transferable.
2. It does not provide legal rights without registration.
3. It may lead to monopoly.

Summary

The term "moral rights" is a translation of the French term "droit moral," and refers not to "morals" as advocated by the religious right, but rather to the ability of authors to control the eventual fate of their works. An author is said to have the "moral right" to control his or her work. The concept of moral rights thus relies on the connection between an author and his/her creation. Authors may seek moral rights protection from state moral rights laws

Moral rights protection varies around the world. For example, in the United States, moral rights protection is at a lower level that in many European Union countries. France, by way of comparison, has one of the most protective moral rights regimes in the world

The main features of related rights in relation to registration, matters covered, duration of protection, rights covered, requirements, authorization, enforcement and infringement have been discussed.

Comparison of moral rights and mask work rights has been presented in tabular form. Advantages and disadvantages of moral rights have been outlined.

Glossary

Disclosure: Right the right to withhold publication of an image or broadcast

False attribution Right: The right not to have a work falsely attributed to them.

Integrity Right: The right to prevent derogatory treatment of their work.

Moral Rights: Moral rights are a special extension to copyright that gives the originator of a copyrighted work rights over its use

Paternity Right: The right to be acknowledged as the author.

Questions

1. **What do you mean by moral rights? Describe their significance.**
2. **Describe briefly the advantages and disadvantages of moral rights**
3. **Describe moral rights on the basis of following points.**
   a. Rights protected
   b. Registration
   c. Duration of protection
   d. Authorization
4. **Describe briefly the main features of moral rights.**
5. **Give a comparison of moral rights and mask work rights in tabular form.**
6. **Write short notes on the following:**
   a. Moral rights
   b. Infringement
   c. Copyrights
   d. Rights to Integrity
7. **Describe briefly the rights covered under moral rights.**
8. **Explain briefly the following:**
   a. Right to Integrity
   b. Right to attribution
9. **What legal actions can be taken against infringement of moral rights?**
10. **Describe briefly the moral rights granted to authors in Canada.**
11. **Explain briefly the moral rights available to authors in United Kingdom.**

Chapter 19

Supplementary Protection Certificates

Introduction

A supplementary protection certificate (SPC) is a sui-generic form of intellectual property right [legal protection] that extends the protection conferred by a patent [called basic patent] in respect of medicinal products, such as drugs, and plant protection products, such as insecticides, and herbicides.

A pharmaceutical or plant protection product must obtain: (i) regulatory approval, (ii) a marketing authorization, and (iii) or product license before its sale in the market. The grant of such an official authorization does often take years because of lengthy testing procedure. Due to the long term procedure for an authorization a patent proprietor may lose valuable time, in which he would have been able to market his product. The SPC compensates for the delay in authorization.

It is estimated that more than 8,000 SPCs for medicinal and plant protection products have been filed in Europe between 1991 and 2003. Main points related to supplementary protection certificate are given below:

1. It is in use in European Union member countries.
2. It is a sui-generis, patent like, intellectual property right.
3. The SPC comes into force only after the expiry of corresponding patent.

Duration: It has maximum life time of 5 years when granted after 10 years after filing date of a patent. The term would be less than 5 years if it is granted after 5 years from the date of filing the corresponding patent.

Conditions of Issue: It is issued against a patent generally immediately on expiry of the term of the patent. It cannot be issued before 5 years from filing date of corresponding patent.

A separate application is to be filed for registration in each country.

It extends the protection conferred by the basic patent beyond the term of that patent but in respect of medicinal or plant protection products covered by authorization. It does not extend the term of the patent itself.

A separate certificate is issued for each product, even if they are protected by the same basic patent.

Main Features

The main features of Supplementary Protection Certificate [SPC] in relation to registration, matters covered, duration of protection, rights covered, requirements, authorization, enforcement and infringement are briefly discussed as follows.

1. **Registration:** The registration is required which is issued on national basis. The application is to be submitted in the prescribed application form in the patent office. If the application does not meet the prescribed conditions, the applicant is asked to eliminate the deficiencies within an appropriate term (about two months). If the deficiencies are not eliminated within the term set, the application shall be rejected. Generally, the above mentioned term is extendable. Thus the registration process consists of following steps:
    a. **Filing of application**: The application complete in all respects is submitted in the patent office along with prescribed fee.
    b. **Examination of application**: The application is examined to check whether it meets requirement for grant of supplementary protection certificate. The application is examined to ensure the following things
    c. The application is accompanied by the prescribed fee,
    d. The application has been filed within the prescribed period,
    e. The application contains the required particulars and documents, and
    f. The basic patent is in force
    g. **Issue of Certificate:** If the application meets the desired requirement, the SPC is issued for a specific period for which annual maintenance fee is to be paid every year.
2. **Matters Covered:** In European Union member countries, a **supplementary protection certificate** (short certificate) is a *sui-generis*, patent-like intellectual property right. This type of right is available for medicinal products such as drugs and plant protection products, such as insecticides fungicides and herbicides. The certificate is granted for the products which constitute (i) the active ingradient or combination of active ingradients of a medicinal product or (ii) the active substance or combination of active substances of a plant protection product.
3. **Duration of Protection:** The maximum life-span of the Supplementary Protection Certificate is five years from its effective date. The term of the certificate may not exceed five years. The term of an SPC depends on the date of issuance of the first market authorisation.

a. If the first market authorisation is issued less than five years after the filing date of the corresponding patent, no SPC is granted.

b. If the first market authorisation is issued more than five years but less than ten years after the filing date of the corresponding patent, an SPC is granted for a term corresponding to the period elapsed between the five-year point and the market authorisation issuance date.

c. If the first market authorisation is issued more than ten years after the filing date of the corresponding patent, an SPC is granted for a five-year term.

d. The application for the certificate has to be filed within six months of the date on which the patent is granted. The supplementary Protection Certificate extends the life-span of a basic patent up to 5 years.

4. **Rights:** A certificate may only be grantedto the holder of the basic patent or his successor in title.The certificate confers the same rights as conferred by the basic patent and is subject to the same limitations and the same obligations.

5. **Requirements:** The following conditions should be fulfilled for grant of a supplementary protection certificates.

   The basic patent protecting the product should in force,

   The product has not already the subject of a certificate,

   A valid authorization has been grantedto place the product on the market.

   The authorization should be the first authorization to place the product on the market.

6. **Fees:** The prescribed processing fee is deposited along with the application for a supplementary Protection Certificate. An annual fee is payable for each year of the effective period of the certificate. Any period of less than 12 months is considered as a whole year. For instance, an effective period of 3 years and 6 months will require the payment of 4 years' annual fees. Annual fees are paid for maintaining a supplementary protection certificate. The yearly fees are paid every year three months before the due date and up to six months after the due date. If the yearly fees are paid after the due date, then an additional fee of 20 per cent is due with the yearly fee. The annual fees for a certificate are to be paid from the first year.

7. **Transfer:** The owner of the supplementary protection certificate has the power to transfer the right of certificate to any other person if he so desires. In other words, the owner or inventor may transfer ownership to any one. Employee often assign such rights to their employers as part of their employment contracts.

8. **Enforcement:** A supplementary protection certificate comes into force only after the expiry of a basic patent. In other words, it enters into force when the related patent expires. The followings are also required before enforcement.

(i) The basic patent has not previouslt lapsed or been revoked, and

(ii The annual fees are paid in time.

9. **Validity:** A Supplementary Protection Certificate does not give protection abroad; it is only effective in the country of registration. Applications for a supplementary protection certificate must be filed on a country-by-country basis. There is no unitary European supplementary protection certificate, but national ones only. The certificate will lapse under following situations.

   (i) At the end of the effective period,

   (ii) If the holder surrenders it,

   (iii) If the annual fees have not been paid in time, and

   (iv) If the product covered is no longer placed on the market.

   A lapsed certificate may be restored when a new authorization to place the product onn the market is granted. If a certificate is declared invalid, the fee for the remaining period will be refunded on written request by the holder.

10. **Infringement:** Like that of a basic patent, supplementary protection certificate provides legal rights against infringement of the protected materials.

    Documents provided in support of an application for a certificateincluding marketing authorizations, will normally be made open to public inspection after they are filed or sent to the office.Copies of any documents which are not treated as confidential will be available upon request as in case of documents relating to patents.

SPC and Moral Rights

There are some similarity and some differences between supplementary protection certificate and moral rights which are presented in Table 19.1.

Table 19.1: Comparison of Supplementary Protection Certificate and Moral Rights

Sl.No.	*Particulars*	*Supplementary Protection Certificate*	*Moral Rights*
1.	Governed by	Sui-generic rights	Sui-generic rights
2.	Registration	Compulsory	Not compulsory
3.	Period of protection	Five years	Till the life of the author
4.	Matters covered	Medicinal products and plant protection products	Copyrighted matters
5.	Validity	Country of registration	Country of registration
6.	Enforcement	After registration	After release of copyrighted work
7.	Requirements	Patented matters	Copy righted material
8.	Transfer	Permitted	Not permitted
9.	Infringement	Possible	Possible

Advantages of SPC

1. Registration is not compulsory.
2. It extends life span of a patent by 5 years.
3. It promotes research work on pharmaceuticals and pesticides.
4. It provides benefits to the owner.

Disadvantages of SPC

1. It has short life span [5 years].
2. It does not provide legal rights without registration.
3. It may lead to monopoly.

Summary

A supplementary protection certificate (SPC) is a special form of intellectual property (IP) that extends the protection of a patented active ingredient or combination of active ingredients present in a pharmaceutical or plant protection product after the expiry of the patent. In other words, a supplementary protection certificate (SPC) is a sui-generic form of intellectual property right [legal protection] that extends the protection conferred by a patent [called basic patent] in respect of medicinal products, such as drugs, and plant protection products, such as insecticides, and herbicides.

The main features of supplementary protection certificate in relation to registration, annual fees, matters covered, duration of protection, rights covered, requirements, transfer, enforcement and infringement have been discussed.

Comparison of Supplementary protection certificate and mask work rights has been presented in tabular form. Advantages and disadvantages of supplementary protection certificate have been outlined.

Glossary

Copyright: Copyright refers to a document which grants exclusive right to the author/creater to publish and sell literary or musical or artistic work.

Patent: It is a legal document issued by a federal government that grants exclusive rights for the production, sale and profit from the invention of a product or process for a specific period of time. Patents also grant the right to prevent others from copying the invention.

Sui-generis: Sui-generis refers to things of their own kind or things with unique characteristics.

Supplementary Protection Certificate: The Supplementary Protection Certificate (SPC) is a *sui-generis* form of legal protection that extends protection conferred by a patent (called "basic patent") in respect of a medicinal or plant protection product.

Questions

1. **Define Supplementary Protection Certificate and describe its significance.**
2. **Describe briefly the advantages and disadvantages of Supplementary Protection Certificate.**
3. **Describe Supplementary Protection Certificate on the basis of following points:**
   a. Registration
   b. Requirements
   c. Duration of protection
   d. Validity
4. **Describe briefly the main features of Supplementary Protection Certificate.**
5. **Give a comparison of Supplementary Protection Certificate and mask work rights in tabular form.**
6. **Write short notes on the following:**
   a. Supplementary Protection Certificate
   b. Sui-generis
   c. Copyrights
   d. Patent
7. **Describe briefly the requirements for supplementary protection certificate.**
8. **Describe briefly the registration procedure of supplementary protection certificate.**
9. **Describe briefly the items covered and duration of protection in relation to supplementary protection certificate.**
10. **Under what circumstances a supplementary protection certificate lapses?**

Chapter 20

Indigenous Property Rights

Introduction

Indigenous Knowledge or Traditional Knowledge [TK] refers to the knowledge, innovations and practices of indigenous people and local communities. In other words, it is the: knowledge gained through tradition from ancestors about plants and animals, medicines, agricultural practices and other matters. Main points about traditional knowledge are listed below:

1. Traditional knowledge can be protected by sui-generic rights
2. Traditional knowledge is also known as indigenous knowledge or local knowledge or traditional ecological knowledge.
3. It covers the matured long-standing traditions and practices of certain regional, indigenous, or local communities. In other words, it is gained from experience gained over the centuries and adapted to local culture and environment.
4. It includes knowledge about plants, animals, indigenous treatment of diseases through herbal medicines, cultural practices,etc.
5. Traditional knowledge is based on teachings and experiences and is orally passed on from generation to generation to some persons only. Earlier there were no written document on traditional knowledge.
6. Some forms of traditional knowledge are expressed through stories, legends, folklore, rituals, songs and even laws.
7. There is difference between local knowledge and traditional knowledge. The former existed from decade to century, whereas the latter existed from millennia.
8. It is collectively owned by the people or natives of an ecosystem or region.
9. Traditional knowledge incorporates knowledge of ecosystem and code of ethics governing appropriate use of environment. This code includes rules and conventions promoting desirable ecosystem relations, human-animal

interactions and even social relationships. A combination of traditional and non-traditional knowledge will form a rich and distinctive understanding of life and the world.

9. The ecological knowledge is composed of both traditional knowledge and experiential knowledge [gained through personal experience].
10. Traditional knowledge is very important almost in every sphere of life. Hence, its protection is very necessary otherwise this valuable knowledge will be gradually lost.

Types of Traditional Knowledge

The traditional knowledge has been classified into six major fields such as traditional knowledge of (i) agriculture, (ii) horticulture, (iii) forestry, (iv) medicines, (v) fisheries and human culture. These are briefly discussed as follows:

1. **Agriculture:** Traditional knowledge related to agriculture includes following aspects.

(i) It includes knowledge of plant and animal genetic resources and their conservation.

(ii) Knowledge of production practices including development of plant varieties and animal breeds.

2. **Horticulture:** Traditional knowledge related to horticulture includes following aspects:
   (i) Knowledge of horticultural plant genetic resources and their conservation.
   (ii) Knowledge of production and improvement of horticultural plant species.
3. **Forestry:** It includes knowledge of timber yielding and other useful forest plants and their conservation.
4. **Medicines:** It include knowledge of medicinal plant species which are used for treating human and animal diseases.
5. **Fisheries:** It includes in depth knowledge about fish genetic resources and their conservation.
6. **Human Culture:** It includes traditional knowledge which is expressed through stories, legends, folklore, rituals, songs and even laws.

   Moreover, local people have better knowledge of environment such as snow, ice, weather, resources and the relationship between things. In other words, local people have better insight into the ecosystem of their region or locality.

Main Features

Main features of traditional knowledge in relation to registration, duration validity, areas covered, transmission, transfer, control, rights protected, bio-piracy and knowledge sharing. are briefly presented below:

1. **Registration:** Registration of traditional knowledge is not essential However, registration will provide legal rights to the people of local communities to protect their traditional knowledge.
2. **Duration:** The traditional knowledge can be protected for indefinite period by the local communities. In other words, there is no limit for conservation of traditional knowledge.
3. **Validity:** The traditional knowledge is general used in the area of its conservation by local people.
4. **Areas Covered:** The traditional knowledge is mainly of practical nature, particularly in such fields as agriculture, horticulture, forestry, fisheries, animals and herbal medicines. It also includes cultural matters such as stories, songs, folklore, proverbs, beliefs, cultural values, rituals, community laws and local language, etc.
5. **Transmission:** The knowledge is directly transmitted from individual to individual. Generally, it is passed on from father to son and mother to daughter generation after generation. Local people or natives possess both collective and individual traditional knowledge. The traditional knowledge is passed on by elders to the young. As the elders die, the some knowledge which has not been passed on is lost.
6. **Control.** The Native people have the rights to prevent use of their traditional knowledge by others without permission.
7. **Rights of T.K.:** The local people or natives, who have conserved the knowledge for centuries together, own the intellectual property rights to their traditional knowledge, even if much of it is yet to be written down. No one has the right to document or use traditional knowledge without permission. In other words, the following things are required for use of traditional knowledge:
   - (i) Prior permission of natives
   - (ii) The source of traditional knowledge should be acknowledged
   - (iii) It should be properly interpreted.
   - (iv) There should respect for traditional knowledge.
8. **Bio-piracy:** The illegal collection of indigenous plants by corporations who patent them for their own use is called bio-piracy. There is bio-piracy of traditional knowledge which needs to be prevented.
9. **Sharing of T.K.:** Local people possess both collective and individual traditional knowledge. Most traditional knowledge is shared among community members. But some traditional knowledge may be specific to an individual. This type of traditional knowledge is kept secret by some people who are engaged in treating some human and animal diseases. Moreover, it depends on the individual to decide whether he wishes to share his knowledge with outside parties or not.
10. **Approach:** Traditional knowledge assumes a holistic view including language, culture, practices, spirituality, mythology, customs and evn

the social organization of local people. It can neither be divided nor separated from the local people who hold it. It is rooted in the spiritual health, culture and language of people. It is a way of life. In other words, it cannot be separated into secular[non-religious] and sacred knowledge.

11. **System:** Traditional knowledge is an authority system. It sets out the rules governing the use of resources, respect and an obligation to share. It is a dynamic, cumulative and stable system. Native people consider it truth.
12. **Adoption:** The traditional knowledge is adopted mainly in local areas where it has been developed. On the other hand, scientific knowledge is used in a larger area. In other words, traditional knowledge has limited practical applications as compared to scientific knowledge.
13. **Basis:** The explanation of traditional knowledge is based on examples, stories and myths, whereas the explanation of scientific knowledge is based on hypothesis, theories and law.

Value of TK and Role of Local People

There is increasing appreciation of the value of traditional knowledge. This knowledge is valuable to the indigenous as well as to the modern industry and agriculture. The role of traditional knowledge is briefly presented below:

1. **Medicines:** Many widely used products such as plant based medicines and cosmetics are derived from traditional knowledge.
2. **Handicrafts:** Handicraft and non-wood forest products are based on TK
3. **Sustainable use:** Most indigenous and local communities live in areas which very rich in plant genetic resources. However, local people have made sustainable use of such resources for thousands of years. They have not exploited the resources.
4. **Conservation of biodiversity:** The local people are rightly called as natural resource managers. They are managing plant and animal genetic resources for thousands of years. Women play a key role in the conservation of plant genetic resources as they are the primary harvesters of medicinal plants seed stocks and small grains.
5. **Special skills and Techniques:** The indigenous people, as on site communities, have special skills and extensive knowledge of local environments. Their knowledge provides valuable information to the global community and useful model for biodiversity policies.

Traditional and Scientific Knowledge

There are several differences between traditional knowledge and scientific knowledge. A brief comparison of both tese is presented in Table 20.1.

Protection of Traditional Knowledge

The role of intellectual property (IP) systems in relation to traditional knowledge (TK), and how to preserve, protect and equitably make use of TK, ha

Table 20.1: Comparison of Traditional knowledge and Scientific Knowledge

Sl.No.	*Particulars*	*Traditional Knowledge*	*Scientific Knowledge*
1.	Assumption as	The truth	Best approximation
2.	Type	Sacred and interconnected	Secular and segregated
3.	Teaching through	Story telling	Formal education
4.	Learning by	Work and experience	Formal education
5.	Format	Oral and visual	Written
6.	Approach	Holistic	Analytical [simpler]
7.	Based on	Intuition and experience	Model or theory or facts
8.	Nature	Subjective	Objective
9.	Transfer	Time taking	Fast
10.	Explanation based on	Examples, stories and myths	Hypothesis, theories and law
11.	Credibility	Collective to local people	To individual inventor
12.	Adoption	In local areas	In entire country

recently received increasing attention at global level. The major issues are food and agriculture, the environment, notably the conservation of biological diversity, health, including traditional medicines, human rights and other Indigenous matters. Thus protection of traditional knowledge has become very important. The protection of traditional knowledge can be of two types, *viz.* (i) defensive protection, and (ii) positive protection. These are explained as follows:

1. **Defensive protection:** This includes those measures which ensure that intellectual property rights over traditional knowledge are not given to parties other than the customary traditional knowledge holders. Some countries and communities are developing TK databases that may be used as evidence of prior art to defeat a claim to a patent on such TK.
2. **Positive protection:** This includes the creation of positive rights in traditional knowledge that empower traditional knowledge holders to protect and promote their traditional knowledge. In some countries, *sui-generis* legislation has been developed specifically to address the positive protection of traditional knowledge. Traditional knowledge providers and users may enter into contractual agreements and/or use existing Intellectual Property systems of protection.
   - (i) For success of any project, program or policy on traditional knowledge, the following aspects should be given due importance.
   - (ii) There should be effective participation of indigenous and local communities in decision making and policy planning;
   - (iii) The traditional knowledge should be given due respect and importance, valued, preserved and properly maintained such as the knowledge relevant to conservation and sustainable use of biodiversity.

(iv) The wider application of traditional knowledge should be made with the approval and involvement of the concerned indigenous and local communities, and

(v) There should be equitable sharing of the benefits arising from the utilization of traditional knowledge.

Many natives do not want to share their traditional knowledge in depth with outsiders because they fear that their traditional knowledge will be extracted, misappropriated and misused without respect to them.

Advantages of TK

1. There is no need of registration.
2. Native people have the right to their traditional knowledge.
3. It will help natives in profit sharing for benefits by users.
4. It is useful in sustainable use of resources.
5. It brings respect and reputation to the local people.
6. It is holistic approach to knowledge.

Disadvantages of TK

1. Non sharing of TK will promote monopoly.
2. It is some times lost due to non-disclosure.
3. It is passed on orally to next generation.
4. It is based on personal experience.
5. Explanations are based on examples, stories and myth.

Summary

The Traditional Knowledge [TK] refers to the knowledge, innovations and practices of indigenous people and local communities. In other words, it is the knowledge gained through tradition from ancestors about plants and animals and agricultural practices and other matters.

Main features of traditional knowledge in relation to registration, duration validity, areas covered, transmission, transfer, control, rights protected, biopiracy and knowledge sharing, etc have been discussed.

The value of traditional knowledge and role of local people have been amply discussed. The protection of traditional knowledge can be of two types, *viz.* (i) defensive protection, and (ii) positive protection. These have been explained. Important points for success of any project on traditional knowledge have been listed.

A comparison of traditional knowledge and scientific knowledge has been presented in a tabular form. Advantages and disadvantages of traditional knowledge and scientific knowledge have been listed.

Glossary

Bio-Piracy: The unauthorized use of genetic resources and the knowledge associated with these genetic resources held by the communities living in the locality, by pharmaceutical companies. In other words, the illegal collection of indigenous plants by corporations who patent them for their own use.

Bio-prospecting: Bio-prospecting is the search for biological resources and related indigenous knowledge primarily for the purpose of commercial exploitation.

Community Knowledge: Knowledge held by communities, characterized by common or communal ownership.

Cultural Diversity: Variety or multiple forms of human social structures, belief systems, and strategies for adapting to situations in different parts of the world.

Cultural Heritage: Monumental remains of cultures such as humankind, the dramatic arts, languages and traditional music, and the informational, spiritual and philosophical systems upon which creations are based.

Customary Law: Law that is customary. There have been requests that customary law may be given some recognition in the protection of traditional knowledge.

Expressions of Folklore: Folklore is a component of the broader term 'traditional knowledge'. 'Expressions of folklore' is often used in association with the system of copyright and related rights.

Intangible Heritage: The practices, representations, expressions, as well as the knowledge and skills, that communities, groups and, in some cases, individuals recognize as part of their cultural heritage.

Indigenous Knowledge: This term refers to knowledge held by communities and peoples that are indigenous. In comparison, traditional knowledge is a broader term as some traditional knowledge may not have the characteristic of indigenous knowledge.

Indigenous Peoples: people whose ancestors inhabited a place or country when persons from another culture or ethnic background arrived on the scene and dominated them through conquest, settlement, or other means and who today live more in conformity with their own social, economic, and cultural customs and traditions than with those of the country of which they now form a part.

Intergenerational Equity: a core proposition is that future generations have a right to an inheritance (capital bequest) sufficient to allow them to generate a level of well-being no less than that of the current generation.

Non-Governmental Organization (NGO): A non-profit making, voluntary, service-oriented/development oriented organization, either for the benefit of members (a grassroots organization) or of other members of the population (an agency).

Prior Informed Consent: Consent that is required for accessing genetic resources or the associated knowledge held by communities living in the locality; requires communities to be informed of all consequence of grant of such permission.

Scientific Knowledge. The knowledge. The knowledge gained through education and scientific readings.

Sui-generis: Sui-generis is a Latin phrase meaning "of its own kind". A sui-generis legal system, for example, is a legal system specifically designed to address the needs and concerns of a particular issue such as Plant Breeders Rights, integrated circuits, moral rights, traditional knowledge, etc.

Sustainable Development: development that meets the needs of the present without compromising the ability of future generations to meet their own needs.

Traditional Knowledge Holders: Refers to persons who create or develop and practice traditional knowledge.

Traditional Ecological Knowledge: Traditional knowledge relating to ecological systems acquired by communities living in close contact.

Traditional Knowledge. The knowledge, innovations and practices of indigenous people and local communities. It is the: knowledge gained through tradition from ancestors about plants and animals and agricultural practices and other matters.

Traditional Knowledge Innovations: Traditional Knowledge strictly refers to knowledge in public domain. TK innovations refer to knowledge that is not in public domain.

Traditional Medicine: Refers to health practices, approaches, knowledge and beliefs incorporating plant, animal and mineral based medicines, spiritual therapies, manual techniques and exercises, applied singularly or in combination to treat, diagnose and prevent illnesses or maintain well-being.

Traditional Resource Rights: A broad usage which include intellectual property rights, but denotes broader 'bundles of rights' including for example, human rights, land rights, religious rights, and cultural property.

Questions

1. **Define traditional Knowledge and describe briefly its main features.**
2. **What is Indigenous knowledge ? Discuss in brief its role in agriculture.**
3. **Describe briefly different categories of traditional knowledge.**
4. **Give the comparison of traditional knowledge and scientific knowledge in tabular form.**
5. **What does the following terms signify?**

   a. Traditional knowledge b. Bio-piracy

   c. Biodiversity d. Bio-prospecting

   e. Scientific Knowledge f. Public domain

6. **Write short notes on the following:**
   a. Traditional knowledge
   b. Scientific knowledge
   c. Land Races
   d. *In-situ* conservation
   e. Traditional medicines
   f. TK Holders
7. **Give full form of the following abbreviations:**
   a. CBD
   b. WIPO
   c. TK
   d. TRIP
8. **Define bio-piracy. How it can be prevented?**
9. **Describe briefly ways and means of preserving traditional knowledge.**
10. **Describe traditional knowledge on the basis of following.**
    a. Profit sharing
    b. Protection
    c. Types of traditional knowledge
    d. Transmission
11. **What are the points that should be given due importance in implementing a project on traditional knowledge in a region or locality?**

Glossary

A

Abandonment: An application for trade-mark or copyright registration may be considered abandoned if the applicant does not take steps to complete the process. An application for Industrial design will be considered abandoned if the applicant fails to reply within the specified time to any report that sets out objections to registration.

Abstract: A brief summary of an invention.

Abuse of Patent: Abusing one's patent rights by failing to make one's invention commercially available.

Advertisement: Publication of a trade-mark application in the *Trade-marks Journal*. Details are published to allow opportunity for challenges to the application (opposition).

Allowance: Acknowledgement by the Trade-marks Office that an application is eligible for registration.

Application: The formal request for registration of intellectual property.

Artistic work: Any visual representation such as a painting, drawing, map, photograph, sculpture, engraving or architectural plan.

Assignment: Transfer of intellectual property rights from the owner to another party.

Author: The creator of an artistic, literary, musical or dramatic work.

B

Berne Copyright Convention: An international treaty extending copyright protection in member countries to nationals of other member countries. Canada is a signatory to this treaty.

Biodiversity: Variety of biological organisms found in the entire world or it the some of genes, species and ecosystem in a region.

Breeders' Exemption: Rights provided to Plant Breeders use protected material as the basis to develop a new variety or for other research work; also called research exemptions or breeders' privilege.

C

Cancopy: A reprography collective that grants licenses to photocopy works and collects fees on behalf of its members.

CBD: Convention on Biological Diversity

Certificate of registration: Official confirmation that your design, copyright or trade-mark has been registered.

Certification marks: Marks identifying goods or services meeting a defined standard (*e.g.*, Wool Mark on clothing).

Claims: The basis for filing an application that allows for the determination of the person entitled to the registration of the mark.

Clearly descriptive: A mark that clearly describes a feature of a ware or service and that is therefore not registrable as a trade-mark.

Coexistence Agreement: A binding agreement whereby two or more traders agree how to operate in the same or similar market to avoid any potential confusion.

Commercial Cultivar: A variety which is used for cultivation on vast area.

Condidate Variety. A variety which is to be ptotected under Plant Variety Protection Act.

Copyright: A state monopoly granted to protect artistic, musical and literary works

Collective: Organization that administers rights granted by the copyright system on behalf of copyright owners who have joined that collective.

Compulsory License. The right granted by the Commissioner of Patents to one or more parties to produce a patented invention. These are granted under certain circumstances as a measure against abuse of patent rights.

Copyrights: Rights that provide protection for literary, artistic, dramatic or musical works, computer programs, and sound recordings.

Copyright Act: The legislation governing copyright in in a country.

Copyright declaration: Formal statement that the applicant is the author, the owner of the work, an assignee or a person to whom an interest in the copyright has been granted by license.

Copyright infringement: Violation of copyright through unauthorized copying or use of a work or other subject matter under copyright.

Copyright Office: The central government office responsible for registering copyrights.

Crown copyright: Copyright of works prepared for or published by the government, *i.e.*, government publications.

D

Database: A collection of data organized for rapid search and retrieval by a computer

Deceptively misdescriptive: A mark that deceptively mis-describes a feature of a ware or service in such a way as to potentially mislead the consumer is not registrable as a trade-mark.

Declaration: Formal statement that you are the proprietor of the design and that, to your knowledge, no one else was using the design when you created it.

Description: A description is a basic requirement of an application for an industrial design. It identifies the features of shape, configuration, pattern or ornament that distinguish the design, and indicates whether the design resides in the entire article or a portion.

Design: The appearance of the whole or part of a product, including any pattern or texture applied to a surface.

Design patent: The term used in the United States to describe a form of protection similar to that of industrial design protection in Canada.

Disclaimer: A statement that the applicant does not claim exclusive rights for a portion of the trade-mark.

Distinctiveness: The variety must be distinguishable in atleast one character from previously available varieties.

Distinguishing guise: The unique shape of a product or package that can be registered as a trade-mark.

Domain Name: The virtual address on the Internet.

Dramatic work: Includes plays, screenplays, scripts, films, videos and choreographic works as well as translations of such works.

Due Diligence: Investigations undertaken to assess the strength of one or more intellectual property rights.

Drawings: Drawings (or photographs) disclose the industrial design and are a basic requirement of a design application.

E

Ecosystem Diversity. Biological variation found in the ecosystems in a country.

EPO (European Patent Office): The pan-European body responsible for granting patents across Europe.

Examination Trade-marks. The process through which the Trade-marks Office determines whether an application for trade-mark is registrable.

Exception: A provision in a copyright law permitting the use of a work by defined user groups without the consent of its creator or without the payment of royalties, conditions that would normally constitute an infringement of copyright. Examples of user groups beneficiaries from exceptions are educational institutions, libraries, museums, archives and persons with a perceptual disability.

Example Variety: In DUS, testing, a variety which is used for comparing a particular trait.

Exotic Variety: A foreign variety which is directly recommended for commercial cultivation.

Extant Variety: All released and notified varieties which have not been protected under plant variety protection act.It includes notified varieties, private varieties, farmers varieties and varieties of common knowledge.

F

Fair dealing: Use of works for purposes of private study, research, criticism, review or news reporting that does not constitute infringement of copyright.

Farmers' Exemption: Legal rights provided to farmers to save, use, exchange, share or sell his farm produce of a protected variety; also called farmers' rights or farmers' privilege.

Farmers' Rights: The legal rights provided to farmers to save, use, sow, replant, exchange, share or sell his farm produce including seed of a variety protected under Plant Variety Protection Act

Farmers variety: A variety developed by farmers and used for cultivation for several years.

Fees: Specific money payable for various services provided by Intellectual Property Office.

First to file: A patent system in which the first inventor to file a patent application for a specific invention is entitled to the patent. In most of the countries, the first person to file has priority over other people claiming rights for the same invention.

G

Genetic Diversity: The variety of genes and genotypes found in a particular species.

Globalization: The process of organizing the whole world into a single integrated marketing unit.

Green Revolution: Quantum jump in the productivity of cereals by evolving high yielding varieties.

I

Improved Cultivars: Varieties developed by scientific method and released for commercial cultivation by central or State variety release committee.

Industrial Designs: The process through which the Industrial Design Office determines whether a design warrants registration. The main objective is t determine whether the design is indeed design subject matter, that it is original, and that the documentation meets the requirements of the *Acts* and *Rules*.

Industrial Design Act: Legislation governing registration and ownership of industrial designs in a country.

Industrial Design Office: The central government office responsible for registration of industrial designs.

Infringement: The unauthorized copying or use of a protected form of intellectual property. It may include unauthorized use of industrial design, trademarks and patents.

Industrial Design: Violation of industrial design rights through unauthorized use of a design.

Integrated Circuit Topography: (a) The three-dimensional configuration of the electronic circuits embodied in integrated circuit products or layout designs. (b) Legal protection against imitation of this configuration.

Intellectual Property: A form of creative work that can be protected through a trademark, patent, copyright, industrial design or integrated circuit topography.

Intellectual Property Office: The government agency responsible for the administration of intellectual property laws.

Intellectual Property Right: Legal rights granted to protect a form of intellectual property for *e.g.* a patent right, design right, trade mark right or copyright.

Intellectual Property Assets: Those intangible assets that derive from intellectual property and that, typically, add value to your business.

Intellectual Property Auditing: Investigations that are undertaken to determine the nature and number of Intellectual Property Rights owned by a business.

Invention: An innovation that is both novel and non-obvious.

L

Land Races: The primitive cultivars which were selected and cultivated by farmers for many generations.

Letter of Consent: Letter from one party, typically the owner of a stronger trade mark, consenting to the use and/or registration of another party's trade mark.

Literary work: Work consisting of text such as novels, poems, song lyrics without music, catalogues, reports, tables as well as translations of such works. It also includes computer programs.

License: Legal agreement granting permission **to any person** to use Intellectual Property Rights in exchange for some consideration or benefit. A license does not change ownership of the copyright.

Licensee: A person other than the owner having the same rights as the owner for use and sale of a product, mark or design.

Licensing a design: Granting of an exclusive license from the owner of the design to another party that allows said party to manufacture, use and sell the design, and to sub-license others to do so.

Licensing an invention: Allowing a business or individual to manufacture and sell an invention, usually in exchange for royalties.

M

Madrid Protocol: An international agreement for the registration of trade marks in a number of countries by means of a single application.

Mask work: A two or three-dimensional layout or topography of an integrated circuit (IC or "chip").

Industrial design: Fee required in order to maintain the rights to an industrial design for a second five-year period.

Patents: Annual fee to keep a patent.

Marking Copyrights: Indicating copyright with a small "c" in a circle (©), the name of the copyright owner and the year of first publication.

Industrial design: For registered design a capital "D" is placed in a circle along with the name, or abbreviation thereof, of the design's proprietor on the object to which the design has been applied, or to its label or packaging.

Trade-marks: Trade-mark owners often indicate their registration through certain symbols, such as R in a circle (registered), TM (trade-mark), SM (service mark).

Modern Cultivars. The currently cultivated high yielding varieties.

Monopoly Right: Another term for Intellectual Property Right; the period during which a patent, design, trade mark or copyright can be exclusively exploited by its owner.

Moral rights: Rights an author retains over the integrity of a work and the right to be named as its author even after sale or transfer of the copyright.

Musical work: Work that consists of music plus lyrics or music only.

N

Nom-de-plume: A name used by a writer instead of his or her real name.

Novelty: The newness of a invention, product, marks, variety etc

O

OHIM (Office for Harmonisation in the Internal Market): The European Community's Trade Mark Registration Authority.

Opposition: The process whereby members of the public may object to the granting of a trade-mark registration if they have valid grounds for doing so.

P

Patents: Unauthorized use or sale of a protected invention.

Patent: A legal right provided to the inventor to exclude others from making, using or selling his invention.

Patent Agent: One fully qualified under the rules of the UK's Chartered Institute of Patent Agents.

Patent Application: A technical document describing in detail an innovation for which a patent is sought.

Patent Attorney: A fully qualified to act before the Patent Office in a country.

Patent Cooperation Treaty: An international treaty providing for standardized filing procedures for foreign patents in the countries that have signed the treaty.

Patent examiner: An official with technical expertise charged with the task of classifying a patent or an application, or of determining whether a patent application fulfills the requirements for a grant of patent.

Patent Office: The patent granting authority and disseminator of patent information.

Patent Office Action: An official written communication by the Patent Office on the merits of an application.

Patent pending: A label sometimes affixed to new products informing others that the inventor has applied for a patent and that legal protection from infringement (including retroactive rights) may be forthcoming.

Performers and producers rights: A term used to indicate rights of performers and sound recording producers to be remunerated when their performances and sound recordings are performed publicly or broadcast. Also referred to as "neighbouring rights".

Performing Rights Society: An organization that administers rights to perform musical works on behalf of composers, lyricists, songwriters and music publishers.

Piracy: An unauthorized duplication of goods protected by intellectual property law (*e.g.* copying software unlawfully).

Plagiarism: Using the work (or part of it) of another person and claiming it as your own.

Posthumous work: A work that is published for the first time (or, for certain types of works, published, performed or delivered in public for the first time) after the author's death.

Popular ariety: A widely grown crop variety.

Preferred Embodiment: An explanation, in a patent application, of the most appropriate and useful practical applications of an invention.

Primitive Cultivars: Same as land races.

Prior art: Industrial Design: The body of registered industrial designs or other published designs including catalogues, manuals and other publications throughout the world relevant to the application for registration of a design.

Private copying: Copying of pre-recorded musical works, performers' performances and sound recordings into a blank medium, such as audio tape or cassette, for personal use.

Private Copying Levy: An amount of money established by the Copyright Board on blank recording media, whether made here or imported, by manufacturers and importers and is then sold in Canada redistributed to eligible composers, lyricists, performers and sound recording makers through their professional associations and collectives.

Prosecution: All the steps involved in completing a patent application.

Proprietor: The proprietor of an industrial design is the owner of the design. This can be the person who created it or who paid to have it created or who bought the design rights from the original owner.

Publication: Making available copies of a work to the public.

Plant breeders' Rights: Legal rights granted to the Plant Breeders/owner for the control over the multiplication and sale of their particular plant variety.

Place of origin: A word or depiction that designates the origin of a product or service and that therefore may not be registered as a trade-mark.

Preliminary search: The search of Trade-marks Office records that one should undertake before submitting an application for trade-mark registration.

Pre-publication verification: A second search of Trade-marks Office records to ensure that no confusing trade-marks exist before a given application is published in the *Trade-marks Journal*.

Prohibited marks: Marks that are specifically prohibited from use in the Trade-marks Act

Proposed use: A statement indicating intention to use a certain trade-mark and how that use will occur.

R

Registration: The granting of formal recognition of a trade-mark, copyright, industrial design, or integrated circuit topography by the Trade-marks, Copyright, and Industrial Design Office respectively.

Registered Patent Agent: A specialist entitled to prepare and prosecute patent applications.

Registered trade-mark: A trade-mark entered on the federal government's Trade-marks Register, which formally recognizes the owner's rights to the mark.

Registrar of Trade-marks: The official responsible for the Trade-marks Office in a country.

Registrar of Copyrights: The official responsible for the Copyright Office in a country.

Registrar of Integrated Circuit Topographies: The official responsible for the Integrated Circuit Topography Office in a country.

Renewal fee: Industrial designs: Fee required in order to renew the rights to an industrial design for a second five-year period.

Trade-marks: Violation of trade-mark rights through unauthorized use of a trade-mark.

Trade-marks: Fee required in order to renew a trade-mark registration for an additional fifteen-year period.

Representative for service: A person appointed by the applicant or registered proprietor to receive documents on his/her behalf.

Rome Convention country: A country that is party to the international convention for the protection of performers, producers of phonograms and broadcasting organizations done at Rome on October 26, 1961.

Royalty: A sum paid to copyright owners for the sale or use of their works, or other subject matter.

S

Search: The act of searching through intellectual property records in order to verify whether a patent, trade-mark or industrial design has been previously filed or registered.

SOCAN: The Society of Composers, Authors and Music Publishers of Canada. SOCAN is the only performing rights society in Canada.

Sound recording: A device which produces sounds, such as cassette, record, or CD player.

Special Order Examination: A request to the Patent Office for early consideration of a patent application.

Specification: Part of the patent application. It includes a detailed description of the invention, claims specifying the aspects of the invention for which protection is sought, and the extent of the protection being sought

Stability: Each generation of the variety should be similar.

Supplementary protection certificate (SPC): A special form of intellectual property (IP) that extends the protection of a patented active ingredient or combination of active ingredients present in a pharmaceutical or plant protection product after the expiry of the patent.

T

Tariff: A standard charge for use of copyrighted works. Usually applies to fees paid by users of musical works and cable companies for the rebroadcast of programs.

Trade-mark: A word, symbol or design, or combination of these, used to distinguish the goods or services of one party from those of another.

Trade secret: Information about a product or process kept secret from competitors.

Trade-mark agent: A a person whose name is entered on the list of trade-mark agents and who is therefore entitled to practice before the Trade-marks Office.

Trade-marks Act: Central government legislation governing trade-mark registration in a country.

Trade-marks Journal: A publication of the Trade-marks Office issued weekly and containing all approved applications and Office rulings.

Trade-marks Office: The central government agency responsible for registering trade-marks in a country

Trade-marks records: The inventory of registered trade-marks and pending applications maintained by the Trade-marks Office.

Trade-marks register: The official listing of registered trade-marks.

Trade name: The name under which a company conducts its business. Trade names may or may not be considered trade-marks, depending on the circumstances.

Traditional Knowledge [TK]: The knowledge, innovations and practices of indigenous people and local communities

U

Uniformity: All plants in a variety should look alike.

Universal Copyright Convention: A country that is party to the Universal Copyright Convention [adopted on September 6, 1952 in Geneva, Switzerland, or to that convention as revised in Paris, France, on July 24, 1971].

V

Variety: A genotype released for commercial cultivation either by State Variety Release Committee or by Central Variety Release Committee and notified by government.

W

WIPO (World Intellectual Property Organization): International body responsible for facilitating the granting of Intellectual Property Rights through out the world.

WTO member country: A Country that is a member of the World Trade Organization as defined in subsection 2(1) of the *World Trade Organization Agreement* Implementation Act.

World Trade Organization [WTO]: An international organization that regulates trade and tariffs between nations in order to ensure that trade flows smoothly, predictable and as freely as possible.

Appendices

Appendix 1: Summary of Intellectual Property Rights

Sl.No.	*Type of IPR*	*Registration*	*Period Years*	*Authorization*	*Requirements*
A.	**Primary Rights**				
1.	Copyrights	Not essential	70	Yes	Originality Fixed expression
2.	Patents	Essential	20	Yes	Novelty Usefulness Non-obviousness
3.	Trademarks	Not essential	10	Yes	Distinctiveness No multiple appearances
4.	Industrial Design	Essential	05	Yes	Novelty Commercial use
5.	Utility Models	Essential	6-10	Yes	Novelty Industrial application
6.	Geographical Indications	Not essential	10	No	Special quality Reputation
7.	Trade Secret	Not essential	Indefinite	Yes	Confidential Commercial use
8.	Related Rights Performers and Database creators	Not essential	15	Yes	Originality Usefulness
	Phonogram producers and Broadcasting	Not essential	20	Yes	Originality Usefulness
	Photographers and designers	Not essential	25	Yes	Originality Usefulness
	Film Producers	Not essential	50	Yes	Originality Usefulness

Contd...

Contd...

Sl.No.	*Type of IPR*	*Registration*	*Period Years*	*Authorization*	*Requirements*
9.	Domain Names	Essential	01	Yes	Unique Internet Web address
10.	Trade Names	Not essential	03	Yes	Novelty Distinctiveness
B.	**Sui-generis Rights**				
11.	Database Rights	Not essential	15	Yes	New creation and Investment
12	Mask Work Rights Registered Unregistered	Not essential	1002	Yes	Originality and Usefulness
13	Plant Breeders' Rights Annual Plants Perennial Plants	Essential	1518	Yes	Novelty Distinctiveness Uniformity Stability
14	Moral Rights	Not essential	Lifelong	Yes	Linked with copyrights
15	SPC	Essential	05	Yes	After expiry of basic patent
16	Traditional Knowledge	Not essential	Indefinite	No	Large scale usefulness

Appendix 2: Address of Patent, Copyright and Trademark Offices in India

Sl.No.	*Name of IPR*	*Head Office/Branch Office*	*Location*
1	Patent	Head Office	Nizam Palace, Kolkata.
		Branch Unit	Karol Bagh, New Delhi.
		Branch Office	Parel West, Mumbai.
		Branch Office	Wallajah Road, Chennai.
		Patent Information System	CGO Complex, Seminary Hills, Nagpur.
2	Patents, Designs and Trademarks	Controller General of Patents, Designs and Trademarks	Maharishi Kurvey Road, Mumbai.
3	Copyrights	Registrar of Copyrights	Ministry of HRD, Kasturba Gandhi Marg, New Delhi.
4	Trademarks	Head Office	Old Central Building, M.K. Road, Mumbai.
		Branch Office	Ashram Road, Ahmedabad.
		Branch Office	Chowringhee Square, Kolkata.
		Branch Office	Okhla Industrial Estate, New Delhi.
		Branch Office	Basant Nagar, Chennai

Appendix 3: Country Code Top Level Domain [ccTLD] Extensions for some Selected Countries

Sl.No.	Country	Domain Name	Sl.No.	Country	Domain Name
1.	Austria	.at	26.	Myanmar	.mm
2.	Australia	.au	27.	Nepal	.np
3.	Argentina	.ar	28.	Netherlands	.nl
4.	Bangladesh	.bd	29.	New Zealand	.nz
5.	Belgium	.be	30.	Nigeria	.ng
6.	Brazil	.br	31.	Norway	.no
7.	Cambodia	.kh	32.	Pakistan	.pk
8.	Canada	.ca	33.	Peru	.pe
9.	China	.cn	34.	Philippines	.ph
10.	Colombia	.co	35.	Poland	.pl
11.	Cuba	.cu	36.	Saudi Arabia	.sa
12.	Denmark	.dk	37.	South Africa	.za
13.	Egypt	.eg	38.	Spain	.es
14.	Finland	.fi	39.	Sri Lanka	.lk
15.	France	.fr	40.	Singapore	.sg
16.	Germany	.de	41.	Sweden	.sd
17.	Greece	.gr	42.	Switzerland	.ch
18.	Hungry	.hu	43.	Thailand	.th
19.	India	.in	44.	Turkey	.tr
20.	Indonesia	.id	45.	Uganda	.ug
21.	Israel	.il	46.	UAE	.ue
22.	Italy	.it	47.	United Kingdom	.uk
23.	Japan	.jp	48.	USA	.us
24.	Malaysia	.my	49.	Vietnam	.vn
25.	Mexico	.mx	50.	Zambia	.zm

References

Alexander I., Poltorak and Paul, J. L. 2002. Essentials of Intellectual Property. John Wiley and Sons.

Alexander I., Poltorak and Paul, J. L. 2004. Essentials of Licensing Intellectual Property. John Wiley and Sons.

Amrita Narlikar 2005. The World Trade Organization. A Very Short Introduction. Oxford Uni. Press, USA.

Andrew, H. J., Phill, G. and Terri, R. G. 2006. Plant Diversity.

Arthur, R. M. and Michael, H. D. 2000. Intellectual Property: Patents, Trademarks and Copyrights [Nutshell Series]. West Group Publishing.

Bhagwati, J. 2005 In Defense of Globalization. Oxford Uni. Press, USA.

Carlotte Fiell and Peter Fiell 2000. Industrial Design A-Z. Taschen.

Carrie Russell 2004. Complete Copyright. An Everyday Guide for Librarians. American Library Association

Clarisa, L. (ed) 2000. Intellectual Property Rights in Emerging Markets. Aei Pr.

David, J. Moser 2001. Music Copyright for the New Millenium. Artist pro. 1st edn.

David Pressman. 2005. Patent It Yourself. 11th edn. NOLO.

Donald,A Norman 2002. The Design of Everyday Things. Basic Books.

Gary, P. Sampson 2001. The Role of the World Trade Organization in Global Governance. UN Uni. Press.

Gert, W. R., Paul, V. K. and Bart, K. M. E. 2006. European Community Plant Variety Protection. Oxford Uni. Press.

Glenn Adamson 2003. Industrial Strength Design: How Brooks Stevens Shaped Your World ?. The MIT Press.

Henri Charmasson 2004. Patents, Copyrights and Trademarks for Dummies. For Dummies

Henry, R. J. 2004. Plant Diversity and Evolution: Genotypic and Phenotypic Variation in Higher Plants. Southern Cross Uni., Australia.

Jessica Litman 2006. Digital Copyrights. Prometheus Books.

Jill Gilbert 2004. The Entrepreneur's Guide to Patents, Copyrights, Trademarks, Trade secrets and Licensing. Berkley Trade.

Joanna, R.J. 1997. Merchandising Intellectual Property Rights. John Wiley and Sons

John Heskett 1985. Industrial Design: World of Art. Thames and Hudson.

Joseph, E. S 2006. Making Globalization Work. W. W. Norton.

Laurence R. Helfer 2005. Intellectual Property Rights in Plant Varieties: International Legal Regimes and Policy Options. FAO-UN.

Lee Wilson 1998. Trademark Guide. Allworth Press.

Lynn, H. and Cheryl, D. C. 2004. Design Secrets: Products 2: 50 Real Life Projects Uncovered. Rockport Publishers.

Mitsuo, M., Thomas, J. S. and Patros, C. M. 2006. The World Trade Organization: Law, Practice and Policy. Oxford Uni. Press, USA.

Patterson, L. R. and Stanley, W. L. 1991. The Nature of Copyright: A Law of Users Rights. Universal Georgia Press.

Paul Gouldstein 2004. Copyrights, Patents, Trademarks and related State Doctorines Foundation Press.

Peter Naray 2001. Russia and the World Trade Organization. Palgrave Macmillen

Raymond Loewy 2000. Industrial Design. Overlook TP.

Richards, S. 2006. Patent, copyright and Trademark. An Intellectual Property Desk Reference. Amazon Book.

Richard, W. Stim 1999. Copyright Law. Thompson Delmar Learning 1st Edn.

Robert, A. G. and Jane, C. G. 2002. Copyright: Cases and Materials. Foundation Press, New York.

Robert C. Lind 2000. Copyright Law. Carolina Academic Press.

Robert, W. H. 2005. Intellectual Property Rights in Frontier Industries. Aei Pr.

Singh, Phundan 2010. IPR and Plant Breeders' Rights. New Vishal Publications New Delhi.

Siva Vaidhyanathan 2003. Copyrights and Copy wrongs: The Rise of Intellectual Property and How it Threatens Creativity. New York Uni. Press.

Stephen Elias 2003. Trademark:: Legal Care for your Business and Product Name NOLO 6th edn.

Stephen Fishman 2001. Copyright Your Software. 3rd edn, NOLO.

Stephen Fishman 2006. Copyright Hand Book. What Every Writer Needs to Know NOLO.

Books by the Same Author

A. Plant Breeding

1. Essentials of Plant Breeding
2. Plant Breeding: Molecular and New Approaches
3. Molecular Plant Breeding
4. Principles of Seed Technology
5. Seed Technology: At a Glance
6. Objective Seed Technology
7. Practical in Crop Breeding
8. Practical and Numerical in Plant Breeding
9. Numerical Problems in Plant Breeding and Genetics
10. Objective Science of Plant Breeding
11. Plant Breeding At a Glance
12. Plant Breeding [For Under Graduate Students]
13. Principles of Plant Breeding
14. Cotton Breeding
15. Heterosis Breeding in Cotton
16. Breeding Hybrid Cotton
17. Breeding Transgenic Bt. Cotton
18. Cotton Improvement in India
19. Glossary of Plant Breeding and Genetics
20. IPR and Plant Breeders' Rights [Subjective]
21. IPR and Plant Breeders' Rights [Objective]
22. IPR and Plant Breeders' Rights [At a Glance]
23. Introduction to Intellectual Property Rights

24. Elements of Baby Corn
25. Breeding Crop Plants for Stress Resistance
26. Plant Breeding: Related Legislations

B. Genetics

27. Elements of Genetics
28. Genetics
29. Principles of Genetics [For UG Students]
30. Plant Genetics
31. Cotton Genetics
32. Objective Genetics
33. Genetics At a Glance
34. Objective Genetics and Plant Breeding
35. Objective Genetics and Plant Breeding [Hindi Edn]
36. Glossary cum Dictionary of Genetics and Plant Breeding
37. Molecular Genetics [Objective]
38. Molecular Genetics [At a Glance]
39. Molecular Genetics [Subjective]
40. Genetics and Man
41. Genetics Today

C. Quantitative Genetics

42. Biometrical Techniques in Plant Breeding
43. Application of Biometrical Techniques in Plant Breeding [Hindi Edn]
44. Objective Quantitative Genetics
45. Quantitative Genetics At a Glance
46. Quantitative Genetics

D. Biotechnology

47. Introduction to Biotechnology
48. Plant Biotechnology
49. Principles of Plant Biotechnology
50. Plant Biotechnology: At a Glance
51. Objective Plant Biotechnology

E. Competitive Examination

52. A Guide to Competitive Examinations of Agriculture

www.ingramcontent.com/pod-product-compliance
Lightning Source LLC
Chambersburg PA
CBHW070252260926
51849CB00001BA/234
* 9 7 8 9 3 5 1 3 0 9 6 9 7 *